PROPERTY OF LUCAS COUNTY
OFFICE OF EDUCATION
ALTERNATE LEARNING CENTER

KINEMATICS FOR THE HANDICAPPED

KINEMATICS FOR THE HANDICAPPED

*A Professional
Handbook by
Diane J. Kramer*

AN EXPOSITION-UNIVERSITY BOOK
Exposition Press   New York

FIRST EDITION

© 1973 by Diane J. Kramer

All rights reserved, including the right of reproduction in whole or in part, in any form or by any means, electronic or mechanical, including photocopying, recording, or by any information storage and retrieval system, without permission in writing from the Publisher. Inquiries should be addressed to Exposition Press, Inc., 50 Jericho Turnpike, Jericho, N. Y. 11753

LIBRARY OF CONGRESS CATALOG CARD NUMBER 73-79218

SBN 0-682-47734-6

Manufactured in the United States of America

Published simultaneously in Canada by Transcanada Books

To anyone who knows or would like to know how precious life is.

To my parents; my sister, Dawn; my husband, Joe; and my son, Steven; and all of my friends who have made mine a good life.

I Hear, and I Forget
I See, and I Remember
I Do, and I Understand

Contents

FOREWORD: "Walking" by William N. Hawley — ix

PREFACE — xiii

Part I
INTRODUCTION

1. Challenge — 3
2. Evaluations — 9
3. Teachers — 16

Part II
ANATOMY OF THE HUMAN BODY AND HOW IT MOVES

4. The Framework — 25
5. The Muscles — 31
6. The Central Nervous System — 38

Part III
WHO IS HANDICAPPED?

Introduction — 47

7. Amputations, Bone and Joint Conditions, Congenital Deformities — 49
8. Cardiac Conditions — 52
9. Cerebral Palsy — 54
10. Epilepsy — 59
11. Visual and Auditory Handicaps — 61
12. Emotional Instabilities and Behavior Problems, Gifted Children — 66

13	Mental Retardation	69
14	Neuromuscular Conditions	74
15	Other Disabilities	76

Part IV
KINEMATICS

16	The Classroom	83
17	Passive, Range of Motion Exercises	92
18	Stationary Exercises	96
19	Locomotor Exercises	108
20	Dance Composition	117
21	Therapeutic Exercises	124
22	Jazz Eurythmics	136
23	Ballroom, Social and Folk Dancing	141
24	Demonstrations and Lectures	150

EPILOGUE   155

REFERENCES   157

RECORD INDEX   159

Foreword: "Walking"

TOWARD A THEOLOGY OF FORGIVENESS
by *William N. Hawley*,
Director of the Center
for the Advancement
of Human Values
in Cincinnati, Ohio[*]

Walking is for moving from one place to another—to go to table for breakfast, to climb stairs to bed, to meet a friend, to walk the aisles at the food mart.
Walking is for enjoying from one day to another—to play hopscotch, to play jump the rope, to play cops and robbers, to go dancing through the trees in the park, at parties for weddings and birthdays, with the beloved at nightfall, to meet the returning child.
Walking is for moving from place to place—is for dancing.
Walking is very important for meeting the world, for growing up, for retreating to solitude, for returning to join again, for carrying the day's tasks, for belonging.
Walking makes a big difference in how one's life turns out, in whether one has a life at all, so it's very important for almost everything—like breathing.
Walking can make the difference whether we ever stand up straight or not.
But walking does not come automatically like breathing. It must be learned. It begins while we are still very small, lying alone in our bed quite dependent on the others we see walking.
We see the others walking—the mother who nurses and feeds us, the father who holds and plays with us, the sister or brother or

neighbor who walks toward us, smiles to us, then walks away, leaving us there alone on our back in our bed.

We see others walking, see what they can do with their walking, what we cannot do, and we want to do it too.

So lying on our back or face soon is not enough. We struggle to crawl—then we crawl everywhere we can. Then we try pulling ourselves to stand at table leg, at father's leg, at the stair-steps, by holding to the drapes, by hanging to the tablecloth.

We grunt and push and pull and fall and roll and bump, then try again and keep it up over and over again, and never quit in spite of face-falls and nose bruises—all because we want to be what we feel persons come to be by walking.

Walking is expensive. It is not learned without risking, without falling so hard it hurts, without slipping on the stairs so suddenly it frightens us and our parents, keeps us from trying again—for a while at least.

Walking is not learned without breaking and spoiling and tearing the Dresden doll on the coffee table, the African violets by the window, the linen tablecloth and crystal set for Sunday dinner, our brother's plastic kite, our sister's favorite doll.

Walking is not learned without knocking things over, scaring the cat, tearing the final edition, spilling the wastebasket, spilling this, scratching that—and most of all straining the good humor of everyone.

Walking is expensive for the child too. But walking is more expensive for the mother and father, sister and brother, baby-sitter and grandparents, for walking cannot be learned unless there's freedom enough for the child to try and fail—try and fail, for the child to try and fail—fall and break, break and spoil without punishment.

To walk is to be lovingly free to fall and fail without fine or penalty. To walk is to find courage to risk the dangers of learning to believe one's own legs and body can do it—by being cared for by loving arms and hands, for living trust in us—by being convinced we are valued more than the Dresden doll, more than the African violet, more than the order of the house, more than the need others have for unruffled humor.

Foreword

Walking is not possible unless there is love for us, unless there is love for our wanting to walk, unless we are loved well enough to be convinced that they who feed and house us are free enough within themselves to let us risk with them the dangers of learning to walk.

Some children learn to walk without love, learn somehow to absorb the shocks of falling and failing as well as the shocks of impatience and dissatisfaction and even anger when falling and failing upsets, disturbs, annoys mother and father, brothers and sisters, but that kind of walking never lets a child stand up straight, lets him grow courage for facing criticism and disapproval, lets him be his own person.

Walking is no different from other forms of being—doing, making, saying, acting.

Children must learn from scratch how to do and do well if the life within is to grow and deepen in freedom, in self-confidence, in openness to nature, to other men, to the future.

Walking is no different from learning to see, from learning to feel, learning to distinguish hand from foot, nose from chin, I from me, I from you.

Walking is no different from learning now from then, good from better, hurt from pain, joy from happiness.

Walking is no different from learning to express, to think, to love and receive love. Each way of expressing and communicating that makes us human must be learned from some other person.

And that learning and that purpose cannot happen without forgiveness, cannot come to pass without a loving that loves us, walking or falling, notwithstanding.

---

*Reprinted from *The Christian Home*, June, 1970. Copyright © 1970 by Graded Press.

Preface

Is it fair that we make eggheads and athletes of our children though they be round-shouldered, swaybacked, knock-kneed and flat-footed? Must they be content with crippled knees for the sake of the sport? We teach primary songs and folk dances in kindergarten. We get them started and somewhere during first and second grade we drop the whole thing. By the time they've reached third grade it's "kid stuff."

Suddenly they're in Junior High and they have forgotten how to move. So we hit 'em with the "social graces," the waltz, the fox-trot, the Virginia reel. This is a bummer! They begin to move again but in their own way, doing their own thing. Girls turn to loud music and primitive movements. Boys turn to sports. Adults try to imitate and become physical wrecks and pathetic sights.

It started way back in the first grade because it was decided that maintaining control in the classroom was more important. Academic study should not be disrupted by students fidgeting about.

We chain our young people to their school desks and we expect them to stay attentive. We admonish them for their clumsiness until by the age of ten they are afraid to move freely. Instead they fidget. And we boast that we are preparing them for life?

Part I

INTRODUCTION

Chapter 1

Challenge

Extensive work and research are being done in the field of movement education, but qualified specialists are scarce and school budgets are slim. Administrators, school boards and teachers may be unaware of the desperate need for creative movement experiences. Expensive equipment necessary for training students is left in a dusty corner or hidden in a closet for lack of a teacher who knows its value. No one has been shown how to use it. Strange that this same equipment is so adequately put to use in the presence of a newspaper or magazine photographer.

Your first step toward Kinematic education is to wonder about these things, think about them, relate them to your own school experiences. Why is it that six hours are spent in study hall and only two hours in swimming and physical education? My particular aim is to show you how to inspire the mentally, physically and emotionally handicapped individual to discover the joy of movement, the gift of life, as far as his abilities will allow. Whether you are a parent or a teacher, above all to concentrate on abilities rather than disabilities. (I also hope to show you how to manage the time and space to do just this, but your success may depend on the intelligence of your superiors.)

Identify the needs of each individual, analyze methods to fulfill these needs and cultivate, enrich and strengthen his capacity to adjust to his handicap. Though you are concerned here with the physical adjustment, the psychological adjustment may, in all probability, be most important.

Learn who is handicapped. A child with poor posture and coordination which causes him to be a poor student may grow up to be a poor man without a job. He is even more handicapped by his bad self-image than the blind man who has made the most of what he has.

To identify the problem—to help each student to develop all of his senses—you must develop all of yours. You cannot prevent or correct or help a child adjust to any disability unless you know what to look for. Explore various methods of movement education. Create new and exciting ways of teaching through movement. I challenge you and your students to discover and share this experience.

Experiment by yourself with various musical sounds, songs and rhythms. Perform an activity or movement in slow motion. It takes extreme effort and control to stand or pose perfectly still. Could you perform a combination of movements to the rhythm of a poem? Perform a skipping, running or jumping exercise and you will not only discover how old you have let yourself become, but after short periods of these experiments performed daily you will discover how good it is to feel young and vital again.

The most difficult thing you have done here has been to put away your Puritan conditioning that says using one's body to create an artistic movement is something one does not do. It may be acceptable for "certain people" but let's not make spectacles of ourselves. I cannot be convinced that it is fair of us to keep our children and ourselves tied up in knots for the sake of our dignity. If it is acceptable to use an object—a paintbrush, a violin, the written word—to create a work of art, then the use of nothing but oneself can be even more creative and more fulfilling.

In part 4 you will be introduced to a multitude of activities, movements and dances. You will experience, for example, the benefits of stationary exercises to develop balance, control, poise, strength, coordination and discipline. Music and movement are important for the emotionally disturbed child in need of an outlet for the emotions of anger and resentment. Relaxing exercises are important for calming the hyperkinetic child or a class that is out of control (discipline again). Yelling "SHUT UP" only works a few times. Exhausting exercises are a matter of opinion. Condition yourself and know your limits and your students' capabilities. I can teach a full eight-hour day and in one hour exhaust a class of teenagers including the athletes. Common sense, please!

Small classes are essential, but overcrowded schools cannot

Challenge

accommodate them. Learn to use the space that is given you. Constructive and creative fun is your goal. Better minds in better bodies is your goal. A child may enjoy moving in space that is all his own, moving in his own way at his own speed. At other times he may prefer having a partner or being part of a group.

There will always be the withdrawn child, the sitter. You want to reach him, prod him, bribe him, anything to get him off his backside, to join the class. And having tried and failed, having lain awake nights, you'll want him dismissed from the class. You've decided he isn't going to learn and your confidence and pride in yourself as a teacher have been flattened. STOP! In less than six months he'll be moving. He'll repeat every step that the rest of the class has been practicing repeatedly. Children have a way of coming through for us just when we are giving up.

Without question, the activity is educational although not particularly obvious to the novice or the spectator and dare not be to the students. The mind and body must be coordinated to perform any movement. Every sense is being improved.

You can start preparing yourself now to open up a new world for your student. Get the Big Picture!

LOOK out the window, across the street, at the neighbor's house. What color is it? What color is the roof, the trim? Look at the trees, the leaves moving in the wind, the sky, your room, the furniture. Look at the texture of the sofa, the carpeting, the drapes. Look at the gleam from the smooth wood of the desk that you just dusted with your lemon furniture polish, the television, the book case, the colors of the books on the shelf.

LISTEN to the radio, the cars going by. Are there planes overhead? Listen to the birds, the dog barking. Can you hear a train in the distance? Does your chair squeak when you move? Tap your fingers. Hum a tune. Move your head around in a circle. Hear the crunch at the back of your neck?

TOUCH a piece of fur, your favorite bracelet or pin, an orange, a spring, a satin ribbon, corduroy, velvet, a warm slice of toast, an ice cube. (Shades of the Sensuous Woman!)

THINK about all the things that you have not thought to teach your students before this moment. Read a poem. Write a letter. Balance your checkbook. Balance on one foot. Balance on one foot with your eyes closed. Now, focus your eyes on a spot in front of you at eye level. Keeping the image of it in your mind, close your eyes and balance on one foot. Now that you've picked yourself up, know now that your mind is in control of your body. (Or is your lack of control controlling you?) You are experiencing that which the mentally retarded and otherwise handicapped are being deprived of.

MOVE your fingers, toes, head. Bend your knee. Walk across the room. Hop. How far can you hop on one foot? How far can you lean forward, backward or sideways without losing your balance? These are important things to know. Better to find out now than someday when you are at the top of a ladder and suddenly find yourself at the bottom. All of these simple movements we take so much for granted are an impossibility for victims of cerebral palsy.

Nancy Smith, editor of Jopher's *Spotlight on the Dance*, expresses it in this way:

> Self has emerged, through a strange, distorted rationale, as a kind of vertical, narrow brownstone edifice with the body in the basement, the mind in the parlor, the emotions in the attic, and the spirit or soul hovering somewhere on the rooftop like a television antenna subject to frequent network difficulties. (Intuition is hidden in a closet like an attractive but retarded child.) And the residents of the brownstone—you and I—run frantically from one room to another trying to integrate the housekeeping.

Following are letters from a volunteer and from a school administrator, both of whom seem to have found a way to "integrate the housekeeping."

Being a volunteer for the handicapped has made me see things I would otherwise never have seen. I've seen children who are blind, deaf, mentally retarded and victims of cerebral palsy. One tends to think one child retarded is like all others who are retarded. That, I have found, is far from true. When

Challenge

you are with these children for even just one hour each week, each week you learn something new about them.

We are teaching them and they, in turn, are teaching us. I have discovered so many phases of mental retardation. Some children seem as normal as you and I, yet some do not seem to even hear you or know where they are or what is going on around them. I've been with three of these children who seem to be in a world of their own and found that in all three cases they will respond to march music when they won't respond to anything else.

In most of these children, young or old, there is a great deal of love. To help others as I have been helped by others (I have epilepsy) is why I became a volunteer for the handicapped and I have found great joy each time I'm with them. No one seems odd to them and they have love for everyone. They don't make fun of others as do normal children and adults. Most of them are trying to do their best and are not sitting feeling sorry for themselves as we often do.

I've seen a little girl with cerebral palsy, used to sitting in a wheel chair. But when put on the floor mat she will try so hard just to open her fist and stretch her fingers. To us that is easy, but to her it was very hard. It made me realize that any normal person can do anything if they try hard enough and really want to.

During a class with E.M.R children, one boy wouldn't do a thing but push and shove others. He wouldn't listen or try to do the exercises that we were doing until I stood away from the others and with him. He listened, watched, and ended up performing the dance in perfect rhythm with me. He became serious and quiet, trying hard to repeat the dance and after I told him how well he did when he really tried, he said, "I did do pretty good, didn't I?"

These children ask for so little—just someone to know they are alive and have feelings just as we do. I can do so little to help them. But as long as I know I can be of help to even one child I must try to do the best I can, whenever and wherever I can.

Your program was a tremendous success here at the Newhope Center. It is extremely rewarding to see the children profiting from this experience. However, it is not just the children who have profited but rather the entire staff. Through you, teachers have found that they can be creative in motion and the self-confidence of our teachers has increased. Those that used to say, "Oh, I am just not creative" are beginning to explore. Through you they have gained a new sense of professionalism.

As the director of the program my main concern centers around the quality of our services. Our quality is naturally dependent on the attitude and morale of the staff. Due to the layout of our physical facilities, groups of teachers developed. Each group cooperated with its members; however, there was a lack of cooperation from group to group. Through your program, which has brought almost the entire staff together, a new bond and appreciation has evolved. This workshop could easily be compared to encounter or sensitivity sessions. By being physically active in front of one another we have let our façades fade and laughed at our own inabilities as well as each other's.

As the year moves on I am sure that your efforts will be felt in many facets of our program which have nothing to do with movement.

Once again, my thanks for a mind-expanding experience.

Chapter 2

Evaluations

To evaluate each student's needs and progress you must be knowledgeable as to what abilities might be expected of what age level. Once you have read this chapter I want you to forget it. By that I mean, do not keep referring to it in order to keep your students in line with the charts. Each child is different and no two children will fit completely into any chart. The information will be available should you need it. Should you be required to present an evaluation of your students' progress to the administrator of the school, by all means use these charts as your guide. But do not say in your evaluation that "Johnnie is slow because he cannot jump" or "Johnnie cannot jump because he is slow." He cannot jump. That is a fact. But tomorrow, as soon as you hand in your report, he'll jump, run, skip, and dance and altogether make a liar out of you.

It is important that, from time to time, you keep progress reports on your students. The report should not only tell how each student has progressed in executing various movements but how his success in movement has carried over into his academic work. Set an individual goal for each child so that you both know where you're headed. Be realistic. The clinical experiences and recently published research studies have proven that movement experiences which improve the perceptual-motor capacities of children will play an increasingly important role in education.

To predict and to remediate motor problems at an early age is far more beneficial to the child than to recommend him to remedial programs after he has been subjected to frustrations and failures in the classroom, exclusion from normal play experiences, and social rejection by his peers on the playground.

CHART ONE

Age (in months)

- 1-5   looks at objects. Raises head, lateral head movements. Hand preference begins. Pushes up with arms while lying on stomach.
- 5-9   Grasps objects. Sits alone. Creeps.
- 10-15   Shows objects to others (beginning social behavior). Walks and stands alone.
- 16—+   Names objects. Identifies some body parts.

CHART TWO

Age (in years)

- 2-3   Throws and catches. Jumps. May walk sideways or backwards.
- 3-4   Identifies front, back and side of body. Knows words "right" and "left," but not that they are on opposite sides of the body. May walk on tiptoes, able to walk in a straight line. Runs, stops, starts and turns. Jumps over ten-inch obstacle. Developing ability to crawl, walk, jump forward.
- 4-5   Knows right and left are on opposite sides, but is unable to tell which is which. Balances on preferred foot, arms folded, 3 to 10 seconds. Running, galloping, jumping with proper arm action.
- 6-8   Developing judgments of right and left. Identifies all body parts. Mature catching and throwing patterns. Develops hopping patterns, e.g., 3 left/3 right; 2 right/2 left; 3 left/ 2 right; etc. Skipping.

CHART THREE

A four-year-old child should be able to:

1. Walk perfectly, changing speed, and show a consistent walking speed and rhythm.
2. Balance on one foot with eyes open for 4 seconds or more.

Evaluations

3. Take three jumping steps forward.
4. Jump off a 2-foot high obstacle, and broad jump 2 feet forward from a standing position.
5. Throw a small ball 10 to 15 feet.

CHART FOUR

A five- to six-year old child should be able to:

1. Balance on his preferred foot, with eyes open, arms folded, for at least 4 seconds.
2. Catch a 16-inch rubber ball bounced chest high from distances of 15 feet, 4 out of 5 times.
3. Jump forward and hop forward on one foot 3 consecutive times.
4. Identify body parts, limbs, front, back and sides.
5. Run smoothly in a coordinated manner, with proper use of arms and legs.
6. Jump over a 10-inch-high barrier.

CHART FIVE

A seven-year-old child should be able to:

1. Make all types of left-right identifications, movements and relationships.
2. Throw a ball with proper step (foot opposite throwing arm) and weight shift.
3. Get up from a back-lying position to standing in 1 to 1.5 seconds.
4. Balance on one foot with eyes closed for 5 seconds.
5. Skip, gallop and run in a coordinated manner.

CHART SIX

For a general evaluation of the program on all grade levels the following guides may be used:

Sense of belonging, achievement and responsibility

Do students enter wholeheartedly into activities?

willingly give and receive help?
take pride in something they can do well?
show pleasure in learning new skills?
make constructive suggestions?
show initiative?
pay attention?
play well on their own?
follow the rules?
take "turns" willingly?
show friendliness to others?

CHART SEVEN

Grades

1-2 Runs in group without falling or pushing.
Meets and passes others in a group without bumping.
Forms single and double circles easily.
Walks, skips, sidesteps, gallops, slides.
Does eight or more rhythmic activities well.
Reacts creatively to music.
Does forward somersaults.

3-4 Follows instructions quickly.
Plays in small groups, in stations and in relays.
Differentiates between various rhythms.
Performs well with jumping rope to music.
Follows square dance calls.

5-6 Does eight or more rhythmic activities well and can lead four.
Does jump rope routine to music.
Does several folk dances well.
Does short ball routine to music.
Performs basic fox-trot and waltz steps.

CHART EIGHT

KRAUS-WEBER TEST: It has been established that success in the following tests is directly related to achievement in academic work.

Evaluations

Test 1 Child lies flat on back, hands behind heads, legs outstretched. Hold child's feet to floor. He then pulls up into sitting position. *(Caution:* This must only be used as a test and not as an exercise. Sit-ups with legs straight put unnecessary strain on the lower back.)

Test 2 Child lies on back, hands behind head and knees bent. Hold child's feet to floor. He pulls himself up to sitting position. (Safe to use as an exercise.)

Test 3 Same position, legs straight. He raises feet and legs 10" from floor and holds for 10 seconds. (Safe as exercise.)

Test 4 Child lies face down, hands clasped behind neck, a small pillow under hips. Hold his feet to floor. He is asked to raise his head, shoulders, and chest off the floor and hold for ten seconds. (Use extreme caution with a child who may be swaybacked.)

Test 5 Child lies in the same position as above and raises legs off floor and holds for ten seconds. (Use caution as above.)

Test 6 Child standing, without shoes, bends from hips, keeping legs straight and touches the floor with his fingertips. He should hold the position for three seconds. *(Caution:* Child with hyperextended knees must keep knees relaxed.)

Purpose: The above tests measure minimum physical fitness—gross motor (large muscle) coordination and general posture adjustment.

CHART NINE   (check chart)

A._____Student is attentive
B._____Takes direction quickly
C._____Good self-discipline
D._____Displays good manners
E._____Concentrates well
F._____Socially adjusted
G._____Lack of tension
H._____Knows left and right hands and feet

I._____Good balance
J._____Good coordination
K._____Knows body parts
L._____Knows directions
M._____Knows numbers, colors and alphabet
N._____Can count to music
O._____Can raise arms straight overhead
P._____Can turn in a circle
Q._____Can kneel on one knee
R._____Can perform counter movements (right hand, left foot)
S._____Walks, using proper arm movements
T._____Runs, picking up feet
U._____Marches, in rhythm, picking up knees
V._____Slides, forward, backward and sideways
W._____Jumps without help
X._____Hops well on either foot
Y._____Skips
Z._____Performs 3 or more movements in sequence

The preceding chart can be helpful in determining what your student's needs are. When working with the retarded, the brain-damaged or emotionally disturbed child you may find your course of action in the very beginning with the student's inability to pay attention. In this case it is best to use exercises that can be done quickly, gradually lengthening the time spent on each.

If your problem lies with the student's inability to take direction quickly (item B on Chart Nine), perhaps your directions can be worded differently to be more easily understood. Give only one direction at a time, e.g., "All the boys stand behind the black line and . . ." (they're gone before you can give the next instruction). A more effective way would be, "*When I say go*, all boys *walk* to the black line." (Wait.) "Now stand behind the black line."

Once your students have accomplished A through G (discipline and social and emotional adjustments), you can more quickly proceed to H through R to find their present knowledge

Evaluations

and abilities. Now you know what their needs are. If a student cannot raise his arms straight overhead, ask him to touch the inside of his elbow to his ear. If he cannot kneel on one knee, he probably can kneel on both knees, sitting on his heels, and gradually work up.

Most important to remember is that if a student is unable to perform a certain task, he is learning the "fear of failure." As a teacher you are responsible for giving him tasks in which he can succeed.

Chapter 3

Teachers

DANCE SPECIALISTS

The local dance instructor or the professional performer or choreographer, with an already extensive movement vocabulary, should always be ready and eager for more study, for more experimentation in new ways to move. The change in approach and methods must be from one of stage "projection" to classroom "projection"—from producing to perform to producing to discover and enlighten. Ample time must be spent on examination of theory, acquiring background material concerning handicaps and increasing movement vocabulary with regard to therapeutic movement. Improvement is needed in developing the ability to simplify a combination of movements into single and simple steps. Training in anatomy of the body and anatomy of musical rhythms is of utmost importance.

SPECIAL EDUCATION INSTRUCTORS

The special education teacher is essentially equipped with teaching methods and theory. Time must be spent transferring the teaching of academic subjects, safety rules, manners and grooming from classroom techniques to Kinematic techniques. University courses must be supplemented with the study of movement, anatomy, music, rhythms and dance composition.

The young teacher just out of college may lack the confidence needed to experiment and create. One the other hand, the established teachers may balk at innovative ideas. Schools fortunate enough to have a physical education director on the staff may consider his efforts sufficient and may not wish to bother with supplementary material for their students. Most physical education programs, however, are centered around competitive sports and games. Kinematic classes are geared to the student who needs this type of expression the most and who cannot compete.

PHYSICAL THERAPISTS

The specialty of the physical therapist is a thorough knowledge of anatomy and of therapeutic exercises. Whether or not he is affiliated with a hospital or clinic, he will have access to numerous machines, whirlpool baths and exercise apparatus. Since physical handicaps are his specialty, he may wish to explore exercise therapy as it applies to the mentally retarded or emotionally disturbed persons. Again, as with special education instructors, emphasis must be on increasing movement vocabulary with regard to rhythms. The need for including music and dance movements in their current system is important, and young and old patients alike will appreciate the addition.

PHYSICAL EDUCATION DIRECTORS

If cooperative and eager to learn, physical education directors are the best supporters of the Kinematic method. They already have the proof that only a few of their students excel in sports and games. They are in need of a variety of other activities for the not-so-athletic student. They have most probably been programmed toward a recreational attitude as opposed to the academic "sit still and pay attention" attitude of the classroom teacher, the medical attitude of the therapist or the "show must go on" attitude of the professional dance instructor. Using their wealth of knowledge in the area of recreation and games can be most rewarding as long as everyone is a winner. Dance movements can always be related to athletic movements. Increased knowledge should be in the field of anatomy and therapeutic exercises to eliminate especially knee, back and ankle injuries. Physical educators are becoming discontented with practicing their profession by trial-and-error methods.

MUSIC INSTRUCTORS

Musicians can be of great help with their contribution of a variety of rhythms, teaching students to identify various dance rhythms in an effort to expose them to the classics as well as contemporary tunes. Most of us are forced to dig for acceptable music through trial and error. Perhaps our limited knowledge of music is confined to the current hit song. Our students, even the deaf, are in tune with the latest rhythm. But teachers and students alike must be exposed to both classic and modern composers.

The music teacher, in order to be of help to her students, must first of all learn to move. She can then coordinate her program so that the students can see the relationships between the dances and the music for which they were composed. She may even wish to extend her talents to those students in need of speech therapy by teaching them songs and by the use of rhythm instruments.

THE PARAPROFESSIONAL

Those of you who have little or no background in either theory, music or movement must have a balanced study program using every part of the Kinematic method in addition to individual research and observations of handicapped children in action. You may be a teacher-aide or a volunteer whose job it is to relieve the classroom teacher for more important duties of the academic program.

Learn as much as you can, decide if you enjoy working with preschoolers, elementary school students, high school students or adults, find the kind of music you and your students enjoy most and proceed with all the enthusiasm you can muster. Remember, if you are left in charge of a class, that you and only you are in charge and the discipline of your class is your responsibility.

Knowledge of individuals in the group and group feeling are most necessary. The skill of the teacher as a leader is more important than her skill as a dancer. She must have confidence in herself, be able to motivate others, and watch that her students

Teachers

enjoy and feel successful in the new activity.

Rarely will a group of youngsters fail the leader, if she doesn't fail them. The first step in selling your program is selling yourself, your personality, ability to handle young people, willingness to work hard, dependability, knowledge of movement and the enthusiasm which you transfer to others.

INSTRUCTOR'S SELF-EVALUATION

1. I know my children—their backgrounds, strengths, weaknesses, interests, emotions, and handicaps.
2. I preplan daily, weekly, yearly, and I have alternate plans ready if needed.
3. I make daily activities challenging, worthwhile, and exciting.
4. I am not hampered by a single "lock step" approach, but use many teaching devices, such as movies, slides, records, unit plans, displays, field trips, etc.
5. I have a good understanding of my children's home life; through conferences and home visits I keep a good relationship with parents.
6. In my classroom I have simple, pertinent, and attractive displays.
7. My classroom shows evidence of good housekeeping habits.
8. I am aware of and utilize local resources in teaching community living.
9. I am on constructively cooperative terms with my supervisor and other teachers.
10. I do not let my personal involvements interfere with good teaching.
11. I share my materials, ideas, and techniques with other teachers.
12. I follow recommended objectives and philosophies.
13. I make use of professional and expert services that are available in the school and community.
14. I strive to improve myself in appearance, voice and attitudes.
15. I strive to grow professionally by taking courses, reading professional literature, joining professional organizations,

and attending workshops and conferences.
16. I can say at the end of the day, "I have done right by my pupils, my school, my associates, my profession and myself."
17. I can honestly say, "I am glad I am a teacher."

If you answered enthusiastically "yes" to all seventeen of the above you should have a halo and not a teaching certificate.

A GOOD TEACHER MUST:

1. Be a warm and understanding person.
2. Have a great deal of insight, perceptiveness and sensitivity.
3. Instill trust and sincerity.
4. Be aware of each individual, his movements and his needs.
5. Be flexible, adaptable and objective.
6. Have an abundance of patience.
7. Have a strong mind and body.
8. Be able to take rejection and perhaps failure gracefully.
9. Project a feeling of love, yet be personally detached.
10. Motivate each child to use all of his positive drives in concentrating, observing, knowing, learning, discovering, exploring, creating, analyzing, and responding, as well as communicating appropriately with others.

Your halo must be slipping by now!

ARE YOU FIT TO TEACH?

1. The teacher should know himself well, accept his own shortcomings and determine to overcome them when possible, and be able to recognize in himself when emotions begin to displace reason. In this way, he should not have to work out his own problems at the expense of his students.
2. He should understand his students in terms of their being products of all previous experiences, as well as heredity. Each one is, therefore, different, and hence treating them all alike is frequently futile.
3. The teacher cannot cause growth in his students, but only

influence and direct it to a limited extent. He can remove obstacles, add materials to make greater growth possible, and aid in every way possible to help the child achieve self-realization of his own potentialities.

4. Immaturity in all its forms—lack of knowledge, misconceptions, prejudice, sensitivity, tensions between individuals and groups, and reasonable fears—is the reason for the existence of the teacher as a professional person.

5. Next to mastery of subject matter, the teacher's own attitudes toward students are the most important factors in his success. If he likes them, is consistently firm and patient in applying pressure toward achieving high standards, and can wait patiently for favorable results, his teaching will be successful.

6. What the teacher is and does is more influential on students than anything he may say. It is sometimes a shock to teachers to realize how much their students are concerned with what they do, say, read, wear, enjoy, and, generally, their entire manner and behavior.

7. Discipline is a slow process of transferring authority from without (parents, teachers, law enforcement officers) to the individual's own personality (self-discipline, self-control, maturity). The older the student, the greater the proportion of reason and explanation and self-participation in enforcing discipline. Without outside standards at first, self-discipline is nearly impossible.

8. A permissive attitude coupled with firm discipline is the quickest route to responsibility and self-control, especially when the latter is applied with consistency, kindness, and thoughtfulness.

9. Authoritarian, inflexible, and impersonal attitudes in the teacher encourage and keep alive rebellion, negativism, and hostility in the pupil.

10. The good teacher should have a personal philosophy that will tolerate frustration, and defeat. He works for the long-run goals. He has the habit of reasonable expectation rather than wishful thinking. He has respect for, but does not wor-

ship facts. He can be uneasy without being unhappy. He can tolerate uncertainty without being paralyzed by anxiety. He can show joy and enthusiasm as well as righteous indignation.

HELPFUL HINTS

1. Check equipment and materials needed for activity before class.
2. Wait until all are listening. Speak clearly and adjust vocabulary to level of students.
3. Be enthusiastic, well organized, and know the activity.
4. Teach activities as a whole. After general rules, skills, etc., are learned, teach the fine points.
5. Teach the students that if they become too tired to continue an activity, they should stop and ask someone else to take their place.
6. When activities involve two people meeting as they walk or run, teach them to keep to the right.
7. In teaching right and left, it is helpful to remove everyone's right or left shoe. (Also remove socks to prevent injuries from slippery floors.)
8. Plan so that instructions do not consume so much time that students lose interest.
9. Watch for loss of interest and "kill the activity before it dies."
10. Keep every student as active as possible. This may require several lines, circles or stations.
11. Teach counting off by twos, threes, fours, etc., for efficiency in making various formations for marching activities.
12. To add enthusiasm form circles, lines, etc.; challenge students by counting or timing the length of time it takes.
13. Emphasize taking care of oneself.
14. To hold attention, silence says a lot.
15. Dramatize the wrong way to do something.
16. Lower your voice rather than raise it to compete with the "chatter."

Part II

ANATOMY OF THE HUMAN BODY AND HOW IT MOVES

Chapter 4

The Framework

OUTLINE:
  I. AXIAL SKELETON
     A. Skull
     B. Spine
        1. atlas
        2. axis
        3. cervical vertebrae
        4. thoracic vertebrae
        5. lumbar vertebrae
        6. sacrum
        7. coccyx
     C. Thorax
     D. Pelvis
     E. Clavicle

  II. APPENDICULAR SKELETON
     A. Femur
     B. Tibia and Fibula
     C. Knee
     D. Foot
     E. Humerus
     F. Radius and Ulna
     G. Elbow
     H. Hand and Wrist

AXIAL SKELETON

The skull is divided into two groups of bones. The cranium consists of 8 bones and forms the shape of the head. The face and lower jaw is made up of 14 bones. The bones of the face are firmly joined to the cranium. The lower jaw, hinged to the base of the cranium, is the only bone of the skull that moves. The lower jaw performs two movements, forward (protraction) and backward (retraction). (The tongue also has these two movements.)

The spine is made up of 33 vertebrae, 24 movable or true vertebrae and 9 fixed or false vertebrae. The spine holds you straight and erect and is the main support of your body, yet you can turn your head (rotation), twist your body (rotation), bend forward (extension), bend backward (flexion), and bend sideways (lateral flexion). Each vertebra is separated by an elastic cushion (disc) which serves as a shock absorber.

The atlas, the first vertebra at the top of the spinal column, carries and bears the weight of the skull. The atlas allows the head to make the "yes" movement.

The axis, the second vertebra, projects upward from its "body," forming a pivot for the ringlike atlas to revolve around. The pivot is called the tooth or odontoid process. The axis allows the head to make the "no" movement.

The cervical vertebrae, the top 7 in the neck, are the most flexible. They allow the head to turn in all directions, and the ligament nuchae—a thick elastic band—helps to sustain the weight of the head (and keeps it from falling off when you bend over).

The thoracic vertebrae, the following 12, have the ribs attached to them, forming the chest cavity.

The lumbar vertebrae are the 5 that form the hollow of the back and allow you to bend forward and backward, twist and bend to the side.

The sacrum is made up of 5 "false" vertebrae in the small of the back. When you are doing sit-ups with straight legs you are straining the bones and muscles of the sacrum. So please do sit-ups with your knees bent.

The coccyx is made up of 4 "false" vertebrae and is called the tail bone. It is the least movable of all vertebrae and is wedged between the bones of the hips.

The thorax is like a bird cage with an opening at the top and the bottom. The ribs, oval-shaped strips of bone, are attached to

The Framework

the twelve thoracic vertebrae on each side. They form the thoracic cavity and protect your lungs and heart and give the chest its shape. The ribs start at the spine and all but two (the floating ribs) are attached to the breastbone (sternum). The sternum is a flat bone shaped like a small dagger pointed down. The connection of the ribs to the sternum is made with elastic cartilage and makes it possible for you to expand, raise and lower your chest when breathing.

The pelvis bones are immovable bones that support the upper body and are the axle on which the legs revolve. The 2 bones of the pelvis flare outward, and below the flare is a cavity or socket which receives the head of the thigh bone (femur).

The clavicle (collarbone) is the connecting link between the body and the arm. The clavicle rests on the sternum and extends out to the shoulder. It has movements like the jaw—protraction and retraction. Teachers should be aware that the clavicle is a bone easily broken in young children. This kind of a break is called a "green stick fracture" because the bone behaves like a bough of green wood when forcibly bent.

The outer end of the clavicle is attached to the shoulder blade (scapula). The scapula is a flat, triangular bone with a handlelike bone rising from it. The shoulder blades are not connected to the ribs, but slide over them. In the outer point of the scapula there is a socket which receives the top end of the upper arm bone (humerus). Even more than the hip joint, this ball and socket allows movement of the arm in all directions (circumduction, flexion, extension, abduction, adduction, and rotation).

APPENDICULAR SKELETON

The femur is the strongest bone of all 206 bones. At the end of the neck of the femur, which is bent inward at an angle, is a ball that fits into the socket of the pelvis. This allows the leg to move in all directions (circumduction) so that you can run, jump, kick and dance. No matter what position you assume the thigh will support its share of the burden. The "Y" ligament, resembling an inverted Y, controls the degree of movement possible in the hips and limits the free movement of the hip joint.

The tibia and fibula are the two bones running parallel to each other from the knee to the ankle. Though they are equal in length, it is the tibia alone which makes the joint at the knee. The fibula is attached to the under surface of the wide head of the tibia. At the lower end both bones hang free to form the ankle.

The knee is the hinged joint which makes the movement of the leg smooth and easy. When the knee is relaxed, the leg is allowed to bend, and when it locks the leg is straight. In front of the knee joint is the kneecap (patella), protecting the knee and helping it to bend smoothly. The patella is attached to the tibia by a ligament which, although flexible, is not elastic and cannot stretch. This allows the knee free movement without dislocation.

The foot has 26 bones set in a semicircle, forming a perfect arch. This peculiar structure is the secret of the foot's strength to support all of your weight. Wedged in between the tibia and fibula, the foot can move up (dorsal flexion) and down (plantar flexion) and can point its toes or sole in (inversion) or out (eversion). The motion of the foot is aided by flexible toes, which like fingers, can move and wriggle and clutch.

The humerus hangs from the shoulder blade (scapula) and is the largest bone of the upper body. Its wide head is shaped like a ball facing in toward the body, while its shaft narrows below the head and forms a graceful line to its base, where it fans out abruptly.

The radius and ulna are the two parallel bones of the forearm which make it possible for the hand to turn completely around. The larger of the two bones is the ulna, which tapers toward the bottom on the side of the little finger. On the thumb side is the radius, which with the ulna forms the wrist and the elbow. When the palm of the hand faces up or forward (supination), the ulna and radius are straight. The radius, by gliding over the ulna, makes it possible to turn the palm face downward or backward (pronation). A bone can not twist by itself, but long narrow bones, side by side, can glide over one another, giving flexible action.

The Framework

The elbow is formed at the base of the humerus that receives the head of the ulna. Like the knee, it is a hinge joint that locks when the arm is straight. In the case of a hyperextended elbow, as in the knee, avoid full extension of the arm.

The hand and the wrist form the most useful tool in the world. The wrist, palm and fingers are made up of more than 50 bones. The hand makes a hinge joint with the wrist and is able to move up and down, around and over. The thumb can reach around and touch any of the four fingers. Each finger has three joints, the thumb has two.

What are the bones made of? The most sensitive part of the bone is the thin membrane (periosteum) covering of the bone which contains small blood vessels and nerves. The cushions which cover the bones at the joint (hyaline cartilage) protect the bones from friction. The joint space is filled with a pouch (bursa) which contains a special fluid or lubricant (synovial fluid) and prevents movable joints from rubbing together. The bones are held together and in place by groups of fibers (ligaments) that can stretch but cannot then return to the original length as muscles do.

A bone will not heal if the covering (periosteum) is badly damaged. When a bone is fractured and the two edges are brought together, the affected area is surrounded by clotted blood and lymph. Each edge begins to make new cells and to push them out toward the other end of the break and stops only when the break is entirely healed. A compound fracture is when the bone breaks through the skin. In time, enough calcium is deposited around the broken end to form a thick swelling (callus) and the bone becomes stronger than before. If a bone is out of place, it is called a dislocation. If a ligament is injured, it is called a sprain.

How do bones move? There are five types of joints which allow movement: elastic, found between the vertebrae, ribs and sternum, which serves as padding or shock absorbers and allows limited movement; hinge, found at the elbow, knee, toe and ankle, which allows movement in one direction; ball and socket, found in the hip and shoulder, allowing movement in many direction; pivot, found between the axis and atlas, allowing a turning movement; and gliding, found in the wrist, ankle and toes, allowing up-and-down and side-to-side movements.

There are four types of angular movement: flexion, decreasing the angle between bones; extension, increasing the angle between bones; abduction, movement away from the median plane; and adduction, movement toward the median plane.

Chapter 5

The Muscles

I. MUSCLES OF THE JAW
   A. Temporalis
   B. Masseter
II. MUSCLES OF EXPRESSION
III. MUSCLES OF THE NECK
   A. Sternomastoid
IV. MUSCLES OF THE SHOULDER AND BREAST
   A. Deltoid
   B. Pectoralis major
V. THE RIB MUSCLES
   A. External intercostal
   B. Internal intercostal
VI. THE TRUNK MUSCLES
   A. The abdominal wall
VII. MUSCLES OF THE ARM AND LEG
   A. Biceps
   B. Triceps
   C. Hamstrings

Some authorities list 659 separate muscles while others regard some muscles as portions of adjacent ones. Muscles are divided into two groups: voluntary, those you can move at will; and involuntary, those you cannot control. The 206 bones of the skeleton are completely covered by muscles. The shape and posture maintenance of the body, pushing food through, circulating blood, producing heat, rhythmic respiration, and the motions the body performs, are caused by the muscles.

Bones can move only when they are pulled by a muscle and when it is signaled to do so by the brain. Muscles are made up of bundles of stringy fibers bound together in bunches. The main part is the fleshy part and can expand and contract like a rubber band. Some are connected directly to the bone while other muscles are connected by means of tendon, a tough tissue tightly bound to the bone. Some muscles are connected to the skin and other tissues.

The strongest and thickest tendon in the body is the Achilles tendon, which starts at the middle of the lower leg, runs to the base of the heel, and is about five or six inches long. The calf muscles, along with the Achilles tendon, are responsible for lifting the heel and pointing the foot. Supposedly, long calf muscles are best for jumping, but Rudolf Nureyev disproved this theory. When there is noticeable pain in the lower leg, exercise should be immediately discontinued for at least ten days. When resuming activity, caution should be taken in jumping and standing on tiptoe. Constant and proper stretching avoids "Charley horse" or bunching of calf muscles, which is caused by jerky movements, lack of relaxation between exercises, insufficient "warming-up" prior to exercise sessions and, in women, may be caused by wearing high heels. Students with short or tight Achilles tendons have difficulty in pressing the heels down to the floor and may compensate by acute overrolling of the foot.

Each muscle is fixed at one point (origin), and the other end (insertion) is usually attached to that part of the body which it puts into action. Motion is made possible by the ability of the muscle to contract toward its point of origin. Muscles do not push; they only pull. Flexor muscles bend your bones at the joints, and extensor muscles straighten. An injury to a muscle is called a strain.

Like the bones, each muscle has its job. There are the prime movers, muscles that put the body into action, and the antagonists, muscles that relax as the prime movers contract. The origin of the prime mover is made stable by the fixation muscles that surround the prime mover at its origin so that all of the force is extended on the other attachment (synergist). This prevents unnecessary movement in the performance of an action.

The Muscles

Muscles are arranged in antagonistic pairs, with each muscle in the body having an opponent in the same area. For every muscle that bends a joint, there is one that straightens it and for every muscle that raises a bone, there is one that lowers it. In the arm, the muscle that contracts is called the biceps; the triceps relaxes and the arm is bent. The triceps contracts, the biceps relaxes and the arm returns to its original position.

This harmony of movement occurs all over the body. The muscles never completely relax and, though we may be at rest and motionless, the muscles remain alert. This is what is meant by muscle tone. Diminished tone can be the result of old age, underuse, overuse, paralysis or malnutrition. It is also a result of accumulated waste products. The effect is a tired feeling, an inability to work and physical and mental inefficiency.

Rest and sleep are necessary for restoration of muscle tissue, in order that the renewal of muscle plasma may be accomplished and a store of materials laid up for further use. Massage is beneficial in conditions of nerve-muscle fatigue, to improve circulation while the body is at rest, so that better nutrition is secured and accumulated waste removed without effort by the individual. (The same is true for bruises and sprains.) Always check with a physician before massaging an injured area.

Beyond a reasonable limit overuse is injurious. The muscles may work irregularly or painfully, nutrition declines, wastes accumulate, and the will is no longer in control. "Writer's cramp" is a familiar example. If repair does not keep pace with wear, the muscles not only become tired from overwork and lack of food but also are burdened with poisons and fatigue. Without muscle tone, the jaw would sag, the eyelids would droop, and breathing would stop.

Although the muscles are all bunched together, no one muscle interferes with the work of another. Flat-shaped, round-shaped, long, thin, and diamond-shaped, the muscles cover the body. There are muscles that twist upon themselves, muscles that overlap each other, muscles that go around or over the bones and muscles that are squeezed between other muscles and bones.

The two principal groups of skeletal muscles are:
1. appendicular, muscles of the limbs
2. axial, muscles of the trunk, neck, head and face.

MUSCLES OF THE JAW

Powerful muscles hold the movable jaw to the immovable bones of the head and provide the power needed to bite and chew. The muscles that lift the jaw are the temporalis and masseter muscles. The temporalis has its origin at the cranium on the side of the head. This muscle is wide at the top and narrows to a point as it goes under the cheekbone. Its point is inserted on the lower jaw. The masseter that helps to raise the lower jaw is also at the side of the head, but it has its origin at the cheekbone, and its insertion—a wide one—at the lower jaw. Attached to the bottom of the jaw are the muscles that lower the jaw.

MUSCLES OF EXPRESSION

The muscles of the face are honest. When we try to hide what is in our minds, the muscles of the face give us away. The muscles of the face are small and thin and are scattered all over the face. Their origin is in the bones of the head and their insertion into the flesh of the face. The muscles of the face are hidden by a thin layer of fat. They become pronounced only when we make faces, smile, or frown. Without these muscles, communication would be more difficult. Our facial expressions, very often, are more eloquent than words. Since facial expressions are learned by imitation, the blind must be taught to increase the use of their facial muscles through the sense of touch.

MUSCLES OF THE NECK

Some students will carry the head forward with the neck held straight or tilted down. In this instance, the head is off balance and can affect the balance of the entire body, sometimes resulting in stiffness and pain. The neck in proper alignment with the spine makes good breathing easier. One fourth of all your muscles are in the face and neck. Two of the more important muscles that move the head, the sternomastoid muscles, originate behind the ears at the base of the skull. These muscles run down each side of the neck and meet at its base. They, with the aid of other neck muscles that originate in the back, enable you to turn, bend, lower, raise and hold your head erect. Your safety depends a great deal on the ability of the sternomastoid muscles to turn the head quickly and easily. A stiff neck may result from a sudden sharp movement, sitting in a draft, whiplash, remaining too long in wet clothing or overexercise.

MUSCLES OF THE SHOULDER AND BREAST

In order to raise an arm, a great deal of muscular force is needed. The muscle that does most of the work is the one that bulges at the shoulder, shaped like a triangular shield and named the deltoid, because it resembles the Greek letter Delta. Attached to the outer side of the collarbone and shoulder blade, the deltoid is inserted near the top of the humerus. This muscle raises the arm and thrusts it forward and backward. The shoulder muscle get help from the breast muscle, or pectoralis major, which consists of many flat bands. These bands spread from the collarbone down the outside of the breast bone and across the seventh rib. The outer ends of the muscular bands overlap, twist upon themselves, and are inserted to the top inner side of the humerus. When the arm is raised the muscle untwists. The deltoid lifts the arm; the pectoralis major lowers it.

THE RIB MUSCLES

Working continuously, the muscles of the ribs raise and lower the rib cage, so that we can breathe. Poor postural habits and faulty breathing cause these muscles to work improperly, limiting the movement of the shoulder girdle and arms and altering the balance of the body. These muscles are called the intercostal muscles, and fill the space between the ribs. The external intercostal muscles raise the rib cage, each muscle raising the rib below it. The internal intercostal muscles, which help pull the rib cage down, come up from the hipbone and clutch the rib cage from the front side. When the ribs are raised and lowered by these muscles, the lungs are squeezed and then released. Like the rubber ball on an atomizer, the lungs expel air when they are squeezed and draw in air when they are released. If the thorax is well developed, movement is free, posture improves and breathing becomes more natural, and strength and flexibility in the spinal column, arms and pelvic girdle result. Compensation for stiff chest muscles is a hollow or swayback, a poking head and pronounced protuberance in the torso. Well-controlled and strong muscles in the thorax, shoulder girdle and arms protect the lower back from injury.

THE TRUNK MUSCLES

The trunk is that part of the body which extends from the shoulders to the hips. In the trunk are located the roots of the muscles that move the head, the arms, the legs, and the trunk itself. The largest muscles in the body are located in the back. These muscles hold the spine straight and provide you with great power. In the front of the trunk is a wall of muscle. This abdominal wall is a flexible coat of armor that protects the digestive organs. The job of bending you forward and backward is assigned to the muscles in the front and back of your body. When these muscles are in tune, you look trim and fit. Control of breathing and freedom of leg movements improve with stronger abdominal muscles.

MUSCLES OF THE ARM AND LEG

The long and wiry muscles of the arms and legs (similar to the strings that move a puppet) are concerned with movement only. The muscles of the upper arm bend and raise the forearm. These same muscles also assist in turning the arm. The muscles of the forearm move and turn the hand. The muscles in the thigh move the lower leg, and the muscles in the lower leg move the foot.

Dancing in any form makes a tremendous demand on the foot. It must be strong, supple and as sensitive as the hand, for its use

The Muscles

in positions and movements quite beyond its natural range. The foot has two functions of great importance—support and propulsion. In studying its structure, one must see the foot in terms of the combined static and dynamic demands made upon it. The ligaments bind the bones together and hold them in place, while it is the muscles which maintain the arch. The failure in the strength and tone of these muscles will result in the changed position of the bones, causing stretched ligaments and fallen arches.

There is much variety in types of feet, not only structural but the way the feet are used to support and move the body. Height and weight, the alignment of the legs in relation to the body are important factors. The less weight the feet are required to bear, the more they will be spared of the stresses and strains. The better the body balance, the more easily the weight will fall correctly on the strongest part of the feet. When the foot is taking the weight of the body, it should be holding the floor equally at three points, little toe, big toe and heel. This triangle creates the base giving the ankle and foot the stability and balance it needs. Attention must be paid to proper footwear and correct positioning of the foot.

Walking seems like a very simple action. But walking would be impossible without the cooperation of every muscle of the body. The act of walking is the result of the body's harmony and with every step we take, 300 muscles are in motion. When the left leg moves forward, the right arm moves out to balance it, the shoulders swing to the left, the hips to the right. The left foot touches the ground and the left leg pulls the body forward, the right leg reaches out and every muscle that was tense now relaxes while its mate becomes tense (similar to some married couples). When you are clumsy or awkward it means only that you have violated the laws of the body.

THE TEACHER MUST REALIZE THAT A STUDENT IS NOT A MECHANICAL PUPPET WHOSE LIMBS CAN BE MANIPULATED TO ANY DEGREE.

Chapter 6

The Central Nervous System

One of the most important functions of the brain is the control of movement. There are many areas connected with this control but one of the most important is the "motor area."

The central nervous system consists of the brain, the spinal cord and the nerves extending from the spinal cord. The brain is a switchboard, receiving messages (stimulus), which effects the sense organs or receptors, and answering (response) messages continuously. The spinal cord is the main cable through which incoming and outgoing messages flow. The nerves are the wires that run from the spinal cord to the remote parts of the body, leaving no area without a signaling device.

The incoming messages are carried on a network of sensory nerves or receptors, and the answers are delivered by way of a companion network of motor nerves or effectors. All neurons react to stimulation, and transmit the impulse by disturbing their neighbors along the route of the sensory nerve until the part of the brain that takes care of such matters is informed. The reply is flashed back in the identical manner through a motor nerve to the muscle needing to be activated.

The enlargement of the spinal cord just inside the skull, at the base of the brain, is the medulla. All messages received by the brain must pass through here and all answers must pass through on their way back.

Above and behind the medulla is the cerebellum, the motor area, which makes it possible to walk, dance, play or do anything that requires coordination and balance. The largest section of the brain, controlling sensitivity, intelligence, emotions and behavior, is the cerebrum. Damage to this area can cause a person to be irritable and distractible. It is divided into two sections (hemis-

The Central Nervous System

pheres), and the side that is dominant determines right- or left-handedness. In right-handed persons the left side is dominant. The cerebrum is so large that it not only covers the other sections of the brain but, because it is confined in the limited and rigid hollow of the skull, has formed many folds (convolutions). The ability to think, plan, remember, learn, see, speak, write, make decisions and decide right from wrong, involving voluntary action, is centered in the cerebral hemispheres.

Impulses coming from the sense organs are sensory impulses. After the impulse has crossed the space between the neurons (synapse) in the brain or spinal cord, it becomes a motor impulse until it reaches the ends at the effectors. Motor neurons and sensory neurons are sometimes connected by associative neurons.

The brain keeps in touch with the world by way of the senses, eyes, ears, nose, skin, and tongue. The eye consists of cornea, lens, retina, and optic nerve. There are six muscles which move the eye up and down, and from side to side. Vibrations become sounds to the brain when they pass through the tunnel of the ear. The outer ear is a funnel which collects sounds. The middle ear is like a cave closed by a thin, tight sheet of tissue (eardrum). The inner side of the eardrum contains three small bones, hammer, anvil, and stirrup. Vibration flutters the eardrum, making the hammer beat on the anvil, which vibrates against the stirrup. The inner ear consists of the cochlea and semicircular canals (affecting balance) filled with fluid. Vibrations cause waves to pass over cells and nerve endings, transmitting sounds to the brain.

Odors reach the brain by way of the nose (unless you have a cold). Cold, pressure, pain, moving air currents, heat, stroking or tickling are transmitted to the brain by the nerve endings in the skin. Taste tells the brain whether the food is good or bad. Three thousand taste buds located in the tongue are divided into four types: sweet, sour, bitter and salt.

KINESTHESIA—THE SIXTH SENSE

Kinesthesia is muscle, tendon and joint sense, also called "position sense," which gives us the awareness of the position of the parts of the body without the aid of vision or touch. The receptors in the muscles are responsible for making us aware of the state or position that muscle is in—whether it is contracted or stretched, as well as the intensity of the contraction or stretch.

Kinesthetic sense enables us to know the position of any part of the body and the movement that a part of the body is undergoing, even when that part is moved by someone else.

The student's sense of position in space is dependent on the visual stimuli and the use of the muscles controlling the movements of the eyeball and focusing, the balance organ of the inner ear working in conjunction with the eyes, and the unconscious impressions that reach the nervous system from muscles, tendons, joints and even the soles of the feet in contact with the floor.

With an understanding of anatomy and placement in her mind, the teacher can interpret its language in her own terms, such as "straighten your knees," "lower your shoulders," and "lengthen your back,"—thus enabling her students to develop "kinesthetic sense" correctly and efficiently.

When a student attains correct body placement he should be prepared to apply its principles throughout the rest of his life. He will learn to move with economy of effort, and protect his body from the common injuries that are the result of faulty placement. In both correct posture and placement there is a minimum of stress and strain, use of energy and breathing becomes effortless and efficient.

The best defense against problems of the lower back is for each student to learn and apply his knowledge about postural mechanisms of the body. Training in the prevention of injury is as essential as treatment following an injury. A student with one shoulder higher than the other, or with a hip pulled too far back, or with a mild spinal curvature may feel uncomfortable when his position is corrected. He must constantly correct and overcorrect until his "kinesthetic sense" tells him he is in a good position.

The back which shows some asymmetry with no corresponding spinal abnormality to account for it is very common. It is a failure of the major muscles on either side of the spine to receive and respond equally to the nerve impulses. This failure may be due to

The Central Nervous System

fatigue, rapid growth or even emotional instability.

When the asymmetry is due to some slight difference in the length of the leg, the body can adapt. Knees, because of the tremendous strain in the knee joint, are another vulnerable area in the malaligned body and difficult to rehabilitate once they have been injured. (Any football player with a knee injury can testify to this.)

The feet that roll in or out are responsible for sprained ankles, fallen arches, etc. Most problems occur among students with strong degrees of knock-knees, hyperextended knees and bow legs. In the knock-kneed student there is usually laxness in the joint, which results in instability in the working of muscles and ligaments which control movement. Most knee injuries occur when the knee is bent and the weight of the body is borne by that leg. When it is straight the ligaments pull into place to act as a protection from serious damage.

VOLUNTARY ACTION   The brain must be excited into action. We do not rise from a sitting position without a reason. Whether the reason is important or not matters little to the brain. All voluntary messages are of equal importance. They must be obeyed. The message to stand up may reach the brain through one of the senses. The brain answers through the motor nerves. It knows which of these nerves control the muscles that move the bones that make you rise. When you give the order, the standing muscles are contracted, the sitting muscles lock, the gears turn, the hinges lock, the pulleys pull and you are standing.

If you have performed an action many times, the brain has a record of this action in its vast files. Standing and sitting are actions that you perform constantly, so the brain, in a fraction of a second, gets the picture out of the file, remembers exactly how it was done, and starts the action, although it may be doing a dozen other things at the same time. That is why it is possible to dance and play games in which many actions are performed at once.

REFLEX ACTION   Some messages that the sensory nerves send to the brain are marked urgent. They are the messages of pain. All over the surface of your body are sensory nerves on the lookout for anything that might hurt you. When the point of a pin pricks your finger, the sensory nerve does not wait to see if the brain is doing anything else. It demands the message, stimulus, be answered immediately. The call for help is answered by way of the motor nerves, and the muscles, effectors, of the hand snap the finger away from the pin, response.

These actions are not planned, nor do we have to learn how to do them. They are called reflex actions and are the result of an excited sensory nerve calling a motor nerve to its aid. In many cases the message goes no farther than the spinal cord, where the connection between sensory and motor nerves takes place. Other reflex actions happen through the brain. Bright light, unexpected noises, and odors reach the brain through the senses, and you blink, jump, or bring some other motor response into play. The reflex mechanism of the brain and spinal cord is alert even when you are asleep. It is always protecting and warning you.

IT'S FUN TO BE ALIVE   The ability of the brain to store information, become excited, adapt itself, and make decisions explains why you can walk, talk, and use your body to dance. Sometimes an action is done once, and the brain remembers how to do it. At other times an act must be done over and over before the brain makes a record and puts it in the files. If a new action includes some old ones, the brain sorts out those actions that are familiar and concentrates on that part of the action it must learn. Actions that are repeated often become habits. You do many things at the same time without planning how they should be done.

THE BRAIN AND THE SPINAL CORD   The brain is in complete charge of the vast, complex operation just described. It occupies the entire top of the skull, and guides and controls you. The brain organizes the scattered parts of your body into one unit, makes you aware of things, and incites you to action. The brain also runs the internal machinery of your body without bothering you about it, so that you may devote your time to the things you want to do.

Almost all the sensory and motor nerves that activate the muscles extend from the spinal cord. The cord is a long cablelike structure made up of hundreds of nerve fibers. It is suspended in

The Central Nervous System

the bony canal formed by the long string of vertebrae, bathed by a clear liquid called the spinal fluid. The cord runs up the spine and through the opening at the bottom of the cranium, where it expands and becomes the brain. The brain itself is divided into three sections: the medulla, the cerebellum, and the cerebrum.

THE SYMPATHETIC NERVES The sympathetic nervous system originates outside the medulla. Descending on either side of the spinal cord, it leads to the vital organs. The sympathetic nerves make the heart beat and the lungs breathe, and direct the movements of all the vital organs. These nerves work whether we are aware or not. We have no control over them. They need no one to tell them what to do. Since the day you were born, they have been thinking for themselves. We did not have to teach the lungs how to give us air. The heart knew how to beat from the very beginning. While it lets us control most of our outside actions, the brain very wisely tells us "please don't bother" where such important internal affairs are concerned. After all, we are only human, and we may forget to tell the heart to pump a little faster when our body needs more blood to supply our muscles with more fuel to move our bones so that we can move faster.

The job of the nervous system is to activate or excite muscles. The sympathetic nerves do this with miraculous skill. All your vital and internal organs have layers of muscle of one form or another. The action of a muscle is one of contraction. Every function that the internal organs perform is produced by the continuous contracting and relaxing of muscles. Only the rhythm varies: the timing of the lungs, the heart, or the intestine is different, and the tempo of each of these organs changes under different conditions.

Your body is a machine that does a multitude of things, but you are not quite finished yet. We are more than just machines, because we love and want to be loved, and know why, and dream, and have ideas and faith.

Part III

WHO IS HANDICAPPED?

Introduction

Kinematics is important to anyone, from the child with faulty posture to the severely handicapped who may be confined to a wheel chair. The handicapped person needs to belong, to feel adequate, to earn recognition. The handicapped person is limited in forms of self-expression and achievement.

Even those with severe and multiple handicaps have a right to develop their potentials, however modest, a right to living arrangements that allow them as much independence as they can learn, a right to be treated as individuals, and a right to privacy. No longer are the handicapped to be treated as human vegetables.

It has become evident that the treatment of the child's emotions has a particular bearing on the degree to which he responds to physical therapy. A comprehensive treatment program must involve physical, psychological, social and educational rehabilitation. Our so called "normal" children could use some of the same.

The following chapters will explain several types of disabilities for which the Kinematics program may be adapted.

Chapter 7

Amputations, Bone and Joint Conditions, Congenital Deformities

AMPUTATIONS

The amputee must, first of all, be helped to adjust to being "different." These areas of physical, psychological, social, vocational and recreational adjustments contribute to his sense of security.

Physical adjustments can be made through development of balance. If the amputation is that of a leg, hopping exercises are in order. Posture exercises and walking should be done in front of a mirror. If posture and walking are neglected, scoliosis or lordosis (curvature of the spine) may result. No exercise therapy is to be done without a physician's approval.

Hopefully, the successful physical adjustment will lead to psychological and social adjustments. In order for the psychological adjustment to be complete, he must have a feeling of social acceptance, confidence and self-esteem. The amputee dreads the thought of anyone having to see his stump. Game skills, developed within the limit of his ability, will aid in his social adjustment as will improved competence in movement activities.

Recommended exercises include those which serve to improve walking, strength, speed, endurance, balance, and agility.

WALKING TECHNIQUE: As the weight of the body is shifted onto the good leg, the hip and shoulder on the opposite side are raised. The student must be taught to practice swinging the artificial leg forward rather than sideways using the same length of stride for each leg. In the beginning, to acquire better stability and a natural walking appearance, steps must be short. As a rhythmic gait is achieved the steps may be lengthened.

Musical rhythms can provide the correct pattern.

It is important for the instructor to watch for signs of stump irritation. The student may be overreaching to keep pace with his classmates. Beware of the opposite extreme where the student refuses to try or, having tried and failed, refuses to try again.

In the case of an amputation of the arm or hand, exercises will be needed to improve dexterity in the opposite hand. The body may be thrown off balance so that posture and walking exercises may be required even here.

BONE AND JOINT CONDITIONS

Types of bone and joint conditions include: tuberculosis of the bones and joints, osteomyelitis, and dislocations.

TUBERCULOSIS of the bones and joints most often occurs in the spine (Pott's disease), but may also occur in the hip, knee and other bones and joints. Usually the ends of the bones are attacked. Tuberculosis in another part of of the body or an injury are the primary causes. Symptoms include pain, inflammation, limited joint action, necrotic bone or abscess. Shortening of the leg, stiff joints, atrophy of the muscles or a deformity of the spine may result. Exercises will be prescribed in consideration of and within the limitations of the disability.

OSTEOMYELITIS may be caused by injury or infection. Symptoms include pain, swelling and redness. The inflammation and swelling in the bones cause pressure, cutting off circulation and nutrition to the effected area. Bone tissue is destroyed and pus formed. Results may be severe crippling. Exercises will be prescribed in consideration of and within the limitations of the disability.

DISLOCATIONS are caused by a blow, a fall or unusual pressure that forces the bone out of place, resulting in damage to the ligaments, cartilage, tendons, blood vessels, muscles and nerves. The most frequently affected areas are the shoulder, hip, fingers, toes, jaw, knee, elbow and ankle. Symptoms involve swelling, pain, limited motion, shock or deformity.

Exercises for shoulder dislocations should serve to strengthen

Bone and Joint Conditions

the deltoid muscle by resistive movements done below the shoulder level. The arm should not be allowed to rise above shoulder level, and internal and external rotation must be avoided. Progression of movement should be from passive to active assistive to free active.

In the case of dislocation in the knee joint any jumping, twisting or jarring must be strictly avoided. Exercises in which the knee action is avoided may progress to straight leg raising, knee flexion and extension.

Infectious illnesses in childhood may leave a potential joint trouble that under ordinary circumstances would not show itself, but comes to light under the difficult demands of dance and rhythmic activity.

CONGENITAL DEFORMITIES

A congenital deformity may be the absence of a part or entire extremity (congenital amputation), cleft palate, harelip, club foot, congenital dislocation of a hip (one or both), missing bones, webbed fingers, torticollis (wry neck), or Erb's palsy (paralysis of one or both arms). Any of these conditions may be caused by heredity, failure of the fetus to develop properly, or disease.

Adjustments must be psychological, social and functional. Exercises will serve to increase strength, endurance and coordination. Social activities are of value in developing a feeling of social acceptance.

I would prefer to have classes available for "normal" students and teach *them* acceptance. Better, perhaps would be schools with mixed classes of both "normal" and handicapped so that they learn as they grow and attend school together.

Chapter 8

Cardiac Conditions

The various causes of cardiac conditions include: congenital heart disease (present at birth), rheumatic, syphilitic, subacute bacterial endocarditis, hypertensive heart disease and coronary heart disease. The teacher must be aware of the varying degrees of disability and know the student's tolerance, scheduling frequent rest periods during and following each class.

The physician may prescribe no restriction of physical activity and ordinary activities may be safely tolerated if certain controls are maintained; e.g., if shortness of breath is noted or if the student begins to breathe with his mouth open, activity should cease and the student instructed to rest. A moderately restricted program may be prescribed which eliminates either severe and competitive activities or ordinary activities. If the physician advises a greatly restricted program, the student will be markedly restricted as to ordinary activities or he will be on complete rest, confined to bed or wheel chair.

The first concern of the physician, family, friends and teachers will be to help in the student's physical, psychological, social, recreational and vocational adjustments. The most difficult may be the social adjustment, since the disability is not a visible one, the student's peer group may not understand why this child cannot participate in all activities and he will be unable to maintain group status.

The purposes of any exercise program (with medical approval) should be designed to increase cardiorespiratory efficiency, develop neuromuscular coordination, develop strength and endurance (within the limitations of the condition), and to increase physical work capacity.

Cardiac Conditions

Following are a list of controls which must be maintained throughout the prescribed program:

1. Any movement should be begun slowly, usually in a lying position.
2. Breathing exercises are important, with emphasis on the exhalation during the *effort* phase of each movement.
3. Control with low repetition (2 to 5 times) and rest between each complete movement followed with a longer rest after several repetitions.
4. Any change in body position must be done slowly, e.g., from a lying to sitting to standing position.
5. The teacher must be alert to stop all activity if shortness of breath is noted or if student begins to breathe with his mouth open.
6. The kind of exercise and the number of repetitions will depend on how well the individual's heart reacts and the advice given by the student's physician. It may be necessary to cut the number of repetitions in half, e.g., if the class is doing 10 repetitions of a certain movement, the student with a cardiac condition should only be allowed to do 5.
7. A teacher must, at all times, beware of colds, sore throats and communicable diseases and protect the student from any cold or dampness.
8. Avoid emotional upsets.
9. Above all, short periods of exercise and frequent rest must be maintained.
10. The majority of children with heart defects will limit their own activity. Overprotection is a mistake since it is important to keep their lives as normal as possible.

Chapter 9

Cerebral Palsy

An estimated 10,000 to 25,000 infants are born with cerebral palsy each year. Many light cases, as with retardation, go undiagnosed or unlisted through lack of education and shame which deter the parents and family from seeking help. It is a lifetime problem. Cerebral brain palsy means lack of muscle control or paralysis.

Cerebral palsy may be caused by a number of problems which occur before, during or after the child's birth. Prenatal causes are failure of the brain cells to develop properly, German measles (viral infection), or poor health of the mother in early pregnancy, diabetes, anemia, high blood pressure, lack of oxygen, cerebral hemorrhage, Rh factor, toxoplasmosis—tiny organisms—worms or worm eggs from undercooked pork or rabbit (may also be breathed in through air passages), or A-B-D-blood type incompatability between parents (there are 27 different kinds of Rh factors), or severe metabolic disturbance in the mother during pregnancy.

Natal causes, or those caused at birth are: premature birth, the chief cause being lack of oxygen which may be due to a blocking of the respiratory tract or a prolonged delivery, smoking, kidney infection, the mother being younger than 16 or older than 40, previous premature births, cerebral hemorrhage at birth, awkward birth position, a rapid delivery involving a rapid pressure change, squeezing or other interference with the umbilical cord, or a vitamin K deficiency. (This should confirm anyone's belief that the birth of a healthy baby is truly a miracle.)

Postnatal causes are: accident or injury to the brain, infections involving the central nervous system, e.g., meningitis, encephalitis, brain abscess, or any condition causing a lack of oxygen to the brain.

Cerebral Palsy

Symptoms may develop very slowly and in some cases not even be recognized until a child is a year or two old. Parents, especially of a firstborn infant may be diagnosed as overanxious if they should notice "something wrong" in his development.

EARLY SYMPTOMS: Tension, irritability, may feed poorly or have difficulty in sucking, slow development of muscular control and coordination, listlessness, disinterest in people or objects around him.

Five characteristics may be noted in children with cerebral palsy, depending on the area of the brain affected:

1. Spasticity, persistent and increased muscle tone, stiffness, limited voluntary control, contraction of muscles if they are passively rapidly stretched .The spastic child may demonstrate an introverted personality, fearful and sensitive to strange people and strange situations. The most obvious defects are in the muscles and tendons of the legs and this produces a peculiar type of walking called a scissors gait. He is fearful of falling and dislikes loud noises, preferring the security of familiar surroundings. This child responds best to individual and small group treatments. Exercises and movements must be accurate but slow, resulting in relaxation and lengthening of the spastic muscles and shortening of the antagonists.

2. Athetoid, constant involuntary movements of irregular rhythm, difficulty in controlling hands, speech and swallowing. Instead of the normal wavelike motion by which food placed on the front of the tongue is moved back in the mouth and then down the throat, the wave may be reversed. The child may suffer from malnutrition and his involuntary rejection of food may cause emotional difficulties and distorted relationships with parents. The child suffering from athetosis may be an extrovert, may be quick to anger and may show signs of emotional instability. (So would I.)

The recommended exercise program for this child should begin with relaxed exercise from a relaxed position using rhythm and progressive activities.

3. Ataxic, disturbed sense of balance and coordination. Consider a set of scales, weights on one side of the scales zoom down.

Kinesthetic sense and depth perception are lost. Movement activities should emphasize improvement of balance, coordination, hand skills, walking and rhythm.

4. Rigidity, hypertensity of muscles, lack of elasticity and intermittent rigidity.

5. Tremor, involuntary movements of regular rhythm or pendular pattern, intention tremor (present intermittently) and nonintention tremor (present at all times). Symptoms of tremor and rigidity may both be present. The tremor characteristic of cerebral palsy is not commonly found in children.

There may be only a slight awkwardness in gait or vision may be affected, also hearing, speech, learning ability due to difficulty in communication, spasms, seizures, behavior problems (understandably so) and grimacing due to inability to control facial muscles.

Cerebral palsy may be classified as: monoplegia, affecting one limb; paraplegia, affecting both legs; diplegia, affecting legs and arms; hemiplegia, one sided; triplegia, affecting both legs and one arm; or quadriplegia, affecting both legs and arms (not necessarily equal in all extremities). Cerebral palsy may also be classified as mild, moderate or severe.

The mental ability of the cerebral palsy child may range from retardation to that of superior intelligence. As previously indicated, a child can be mistakenly diagnosed as retarded because of his inability to communicate. A child may be afflicted with cerebral palsy *and* retardation but the two do not automatically go hand in hand. Sadly, the general public, due to lack of understanding, has the mistaken impression that the difficult speech and/or incoordination is a sign of mental dysfunction, which indeed it is not.

Discipline may be a problem with the cerebral palsy child if parents have been either neglectful or overprotective. The movement activities may begin with massage followed by a warm bath. Passive motion may progress to active assistive motion and active motion followed by resistive motion. Avoid stretch reflexes, incoordination and tension. Condition motion will include rhythms and planned movement activity. Confused motion means

Cerebral Palsy

contraction of one muscle or group of muscles by resistance to an unrelated group. Combined motion is action in two or more joints. Reciprocation is simultaneous reverse action of opposite parts or opposing muscle groups of a single joint.

"Motor reeducation" includes both active and passive exercises. The fundamental aim of all types of therapy is to bring out the best, to allow maximum potential and to achieve independence. The student should be taught therapeutic exercises geared to increase their ability to relax and activities which can be performed in a relaxed position. Slant boards permit more oxygen to reach the brain. Also, balance improvement, development of hand dominance through reaching and grasping, instruction in dressing and personal grooming.

Speech and hearing therapy should concentrate on breathing, tongue and mouth exercises. The teacher's goal is to develop the student's ability to use what he has to the maximum, to increase his social development and poise. Each child is an individual in learning capacity as in every other respect. He may be hindered by his inability to concentrate. The path to progress must be custom made.

Most intelligence tests rely heavily on vocabulary and language structure and many others require activities based on language reasoning. For these and other reasons, many authorities feel that present methods of measuring intelligence are meaningless.

The teacher must avoid two stumbling blocks: overprotection and failure to be realistic. The cerebral palsy child must come to accept his limitations, before he can forge ahead to realize his potentials. Evidence being gathered demonstrates that considerable benefits may be brought to a great many cerebral palsy children once regarded as hopeless. They need to learn self-help, manual, creative, recreational, social and educational activities to attain maximum independence through improved physical and psychosocial functioning.

The teacher must aim for muscle reeducation, equalization of muscle imbalance, prevention or correction of deformity, increased range of joint motion, strengthening of weak muscles, improved coordination and training in activities of daily living.

Visuo-motor skills are needed to improve motor planning, gross and fine motor coordination, body schema, and sensory motor establishment of dominance patterns. For beginning walking, stress sidestepping and extending knees, feet kept flat on the floor with the heels down.

Chapter 10

Epilepsy

There are four known classifications of epilepsy. Although the cause is unknown most authorities believe the cause to be brain damage. Grand mal (great illness), lasting 2 to 5 minutes produces strong rhythmic contractions, irregular and noisy breathing, drooling, pale color and loss of bladder control and loss of consciousness following seizures. Before the convulsion, many experience "aura," feelings of impending doom, unpleasant odors, funny sounds, tingling skin and spots before the eyes.

Petit mal (little illness) may produce seizures as frequent as 20 to 100 times a day, and last from 5 to 60 seconds, consisting of twitching eyelids and a blank look. The child often does not remember these episodes.

Jacksonian epilepsy is a convulsion on one side of the body and may be followed by loss of consciousness. Psychomotor epilepsy includes symptoms such as loss of consciousness, periods of abnormal behavior, headache, dizziness, temper tantrums, confusion, mental dullness, amnesia involving aimless wandering, mumbling or dropping articles and either grand mal or petit mal seizures. The child may have warnings of attacks. In young adults, convulsions may be symptomatic of changes in body chemistry, glandular difficulties, rapid growth, menstrual periods, head injuries and tumors.

During a seizure the person should be placed in a lying-down position away from hard or sharp objects. The attack must be allowed to run its course. If possible, place a handkerchief or tongue blade wrapped in soft material between the back teeth and turn him on his side to allow the saliva to flow out of his mouth. At no time should the attempt be made to pry the teeth apart once the seizure has started. Do not restrain or interfere with the individual's movements. It is unnecessary to call a doctor unless the attack is immediately followed by a second seizure, or if the seizure lasts more than ten minutes. For someone helplessly looking on, two or three minutes may seem much longer. When the seizure is over, let the child rest in a comfortable position. Inform the parents that the child has had an attack.

Keep calm. Treat the incident in a matter-of-fact manner. You and his classmates must understand and accept him as an individual. The attitudes of those around him are of vital importance. Do not pamper.

Adjustment for the person with epilepsy is more social than physical. Overprotection or rejection by parents may have caused fear, mother dependency, mental retardation, maladjusted behavior, hyperactivity, distractibility or perceptual difficulty, all of which will have caused greater disturbance than the seizures themselves. As the epileptic enters adolescence, his social and emotional problems increase because of fear, pity, sympathy and rejection. The result may be more frequent seizures. It is important that the epileptic avoid situations involving complicated machinery and mechanical devices, moving objects, high places, cramped quarters or exposure to burns.

All activities involving the use of the body and the mind are antagonists of seizures. The acid produced by the body resulting from activity is of value in contributing to control over seizures. Controls over activities, however, are important in that the child not be allowed to participate in contact sports where there is danger of receiving a blow to the head, highly organized competitive activities and games producing intense emotional reactions and avoiding apparatus such as parallel bars, high horizontal bars, swings or any moving or climbing apparatus. The teacher must explore all avenues of information before jumping head first into the educational process.

Chapter 11

Visual and Auditory Handicaps

In the words of Helen Keller, "The curse of the blind is not blindness, but idleness." Visual handicaps are classified as astigmatism, blurred vision; myopia, nearsightedness; hyperopia, farsightedness; strabismus, cross-eyes; partially sighted and blind. Those persons termed legally blind may have some light or object perception. Causes may be heredity, disease or accident. Hereditary causes include: cataract, retinitis, pigmentosa or atrophy of the optic nerve. Diseases which may cause blindness are: scarlet fever, syphilis, measles, meningitis, mumps, opthalmia neonatorum, tuberculosis or septicemia.

Controls which must be used will depend on the type of blindness. In the case of glaucoma the student must avoid any activity causing pressure. Caution must be taken in performing any activities where the head is lower than the waist, avoiding bumps and falls in the case of a student with a detached retina. Albinos must keep their back to the sun or light.

The physical and social adjustments are most important to the visually handicapped. The therapy program stresses developing memory and increasing auditory, tactile and kinesthetic perception. Because of a slowdown in activities, the student will exhibit a lack of coordination, muscle tone, strength and endurance. His posture is poor because he does not see how he looks to others. He may have to be taught facial expressions. He needs to overcome the fear of falling and bumping into objects by learning conceptions of space, levels and distance. The visually handicapped may partake in any movement activities as long as he has his physician's approval. The teacher's job is to help him toward independence.

Teachers and fellow students should be educated as to the

following items (from the Dayton, Ohio, Goodwill Industries Rehabilitation Center):

1. Always make yourself known; identify yourself with a casual greeting when entering a room occupied by a blind person.
2. Don't talk to a blind person as though he is deaf.
3. In conversation, address the blind person by name if he is the one expected to reply.
4. Read mail to a blind person promptly and refrain from commenting on the content.
5. Help to familiarize the blind with their surroundings; they are at ease when they know where they are.
6. Always leave doors open or shut as you found them; the blind remember them that way.
7. Don't misplace articles belonging to a blind person; it is doubly hard to find things when you cannot see them.
8. In guiding a blind person, never take his arm and push him ahead, instead offer your arm to the blind person. The movement of your body will inform him as to whether or not he should step up or down.
9. Always tell a blind person about changes in the weather.
10. Don't regard blind people as peculiar individuals. As Sir Arthur Pearson said, "They are just ordinary people with very bad sight."
11. Don't discuss blind people before their faces as if they were not present.
12. Never express sympathy for a blind person in his presence.
13. Don't revise your conversation so as to use "hear" instead of "see." Use the word "blind" without hesitation.
14. When a blind person is entering a car or train, going upstairs or about to sit down, he needs only to have his hand placed on some leading object. He can do the rest.
15. Address a blind person directly, not through another person.
16. Don't talk of an "extra sense" or "providential compensation."
17. Don't be patronizing with the blind, be natural in speech and actions.

Visual and Auditory Handicaps

18. When reading a menu to a blind person, read prices as well as food items even when he is your guest. This gives him an indication of the kind of restaurant in which he is eating.
19. When eating with a blind person, never push his food on his plate when it appears to be sliding off without telling him of your action.
20. Tell a blind person you are picking lint from his coat so he will understand your actions.
21. When a blind person leaves the room, going to the elevator, etc. offer assistance if it appears needed.
22. When a person enters a room where there are a number of people present, you can help by suggesting the location of an empty chair. The same is true if a blind person boards a bus on which you are traveling.

Additional notes:

1. If you must leave the room, lead a blind child to a chair or table and place his hand on the object. Do not leave a blind child standing out in space with no reference as to his position.
2. When walking with a blind person, direct him with your words and let him feel that he is leading you.

Classification of auditory handicaps range from the hard of hearing who are able to hear with the help of a hearing aid, to the profoundly deaf. Deafness may be hereditary or may be caused by prenatal conditions, brain disease, scarlet fever, meningitis, measles, typhoid fever or otosclerosis.

Adjustments must be made in social and psychological areas. The deaf may partake in any exercise program with special emphasis on development of balance and rhythm. Problems of speech are improved through the rising and falling inflections and rhythms of dance. Talk to the deaf child. Deaf children must also be made aware that their shuffling feet may disturb hearing persons. Since it is easier for them to imitate your movements than to understand your words get right to the action part of the class and eliminate the bombardment of preliminary instructions.

When you do speak to them, look directly at them so they can read your lips, and enunciate clearly. Do not shout. A drum may be useful in teaching rhythms before the music is introduced. Use anything that makes noise. Turn bass control of the record player to its highest setting. For safety reasons, hearing aids should be removed for movement class.

If a mistake is made, repeat the movement in the wrong way and say, "No," then repeat the movement correctly and say, "Yes." If they are familiar with signs, it will be helpful to learn the signs for, e.g., "slow," "fast," "ready and" (to get them started), or "again." Give them four or eight introductory beats on the drum. When you are pleased with your students you must smile and nod and applaud for them. They cannot hear the tone of your voice.

It is felt that the rhythm with which we dance and play and go about our daily business, is intrinsic rather than extrinsic. That is, we move according to our pulse or our own rhythm of breathing and the music we use is only necessary to make dancing more pleasant for those of us who have hearing. This may explain why the deaf can perform to music they do not hear.

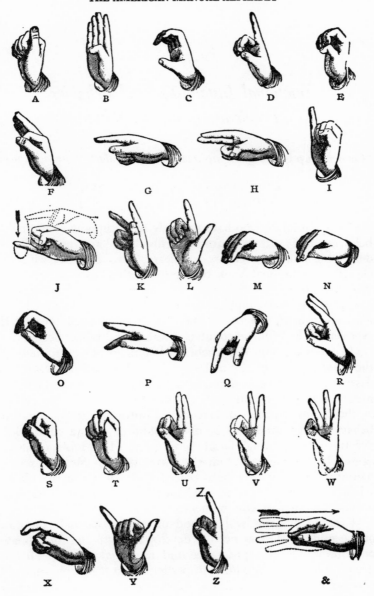

Chapter 12

Emotional Instabilities and Behavior Problems, Gifted Children

"Growing up is like going up a staircase in a dark haunted house."

EMOTIONAL INSTABILITIES AND BEHAVIOR PROBLEMS

Many children, who have seriously handicapping alterations of behavior and certain learning skills but have practically no detectable motor or sensory handicaps, have disorders which are classified as minimal brain damage. In other cases the reverse is true. Unfortunately, this does not guarantee that the child will not have emotional problems since many of the emotional problems are caused, not by damage to the brain itself, but by the experience the child encounters because of his handicap.

In any classroom the teacher is bound to be confronted with the child who is socially maladjusted. Either he is aggressive and destructive or shy and timid, or any number of various labels we may wish to pin on him. Even the fact that he has been "labeled" by the family, teachers or classmates contributes to his abnormal behavior. This child may be obese, blind or deaf, a child with a deformity, or unusual facial appearance. His problem may be caused by excessive or stunted growth, skin disease, or excessive hairiness. He may have been placed in an inappropriate class in school, or he may have an inadequate home, community, neighborhood, or school environment.

Exercise progams will depend on the needs and conditions of the student. He may need to develop strength and functionality or improvement in appearance and improved posture. He may need inspiration or prodding to participate in team activities or to develop superior ability in one particular activity. If the teacher

Emotional Instabilities and Behavior Problems

is on her toes, she may plan success and recognition situations, including moderate use of the child as a helper or leader. The child may have to be protected from situations which tend to cause failure and embarrassment.

Emotional disturbances are found very often in cases where there is no physical or organic reason. Children develop all sorts of emotional problems when they fail to find the security that comes with the knowledge that the adults are in control and that their childlike feelings and emotions will not be permitted to run away with them. We are finding more children who are emotionally damaged by being overly protected and undisciplined than by too much discipline. Parents who are raising children in the same way they were raised may be using the wrong kind of discipline. Times have changed all too rapidly and so must the way we bring up our children.

Perception is the process by which we categorize the information received through sight, hearing, taste, touch, smell, position and movement. The young school-age child may be able to see a *b* and a *d* very clearly as far as the visual process is concerned but he may have difficulties in recognizing the difference between them because of the deficiencies in perceiving position in space. He may have difficulty in learning to work from left to right on the page or be unable to understand the position of things in space. In movement class he will be unaware that his left hand is raised when everyone else is raising their right hand, sometimes even after the error is pointed out.

Unfortunately, if the teacher is unable to recognize the basis for the child's problems, she may continue to drill the child in a particular area when the real source of the difficulty is overlooked. It is helpful to go back to the basic areas of perceptual disturbance (in this case the problem of perceiving position in space) and provide therapeutic exercises.

It is easy to be misled into believing the child's difficulties are purely emotional. Experts working with emotionally disturbed children are becoming aware of the emotional difficulties which arise out of specific defects in the development of the central nervous system.

A handicapped child, as he grows, may become irritable, cranky, shy, withdrawn or defiant, as he realizes that he is different from other children. The failure of the teacher to acknowledge the child's limitations will cause him to become difficult. Realize that the physically handicapped child goes through stages of emotional development just as the nonhandicapped child. Teachers of brain injured children may have a particularly difficult time in determining what is normal and what is an abnormal stage in child development and what is a symptom of brain damage. Frequently symptoms of brain damage may be mistakenly attributed to a willful stubbornness or assumption that the child is spoiled. Likewise, there is always the danger of reading into a child's behavior something that is not there. Let the doctors make the diagnosis.

For the emotionally disturbed, the first form of communication is movement; expressing joy, sorrow, fear, hate, love, hostility, anxiety, passiveness and aggression. With individuals who are unmanageable, antisocial, distractible or with any disruption, the teacher can maintain control with isolation, diversion, body contact or by ignoring him, changing movements rapidly, or taking the student by the hand and gently moving him back into the group.

With autistic, schizophrenic or inhibited neurotic persons there is lack of verbal communication. Movement is very useful in helping these individuals to relate and communicate nonverbally. That which is expressed physically is a measure of his true feelings and emotions.

GIFTED CHILDREN

A gifted child may be classified as exceptionally intelligent (I.Q. 130 or above) or unusually talented in art, music, dance, science or social leadership. Social, emotional or moral adjustments may be indicated with the movement program emphasizing development of strength, endurance and agility. The gifted child needs to succeed socially since he is too advanced to relate to his own peer group and too young to be accepted by older students.

Chapter 13

Mental Retardation

The classification of the mentally retarded differs in various geographic areas. Generally the child termed trainable mentally retarded may have an I.Q. of 50 or below. Children with I.Q.s of 50 to 70 are termed educable mentally retarded. Slow learners will range from 70 to 90 I.Q.

Physical, social, emotional and moral adjustments are necessary for this child to function adequately. The exercise program should contribute to the development of speed, strength, endurance, coordination, and vitality with special emphasis on self-expression.

Mental retardation has been designated by various terms and defined in many ways. Reduced to its simplest form, mental retardation stands for a subnormal intelligence and a reduced capacity for learning.

TYPES OF RETARDATION

1. Mongolism or Down's syndrome. Characteristics of this type of child are a short neck, chubby fingers, a curving little finger, slanting eyes and a straight horizontal line on the palm of the hand. Mongolism is now thought to result from abnormal chromosomal groupings in body cells from the time of the fertilization of the ovum.

2. Exogenous or Brain Damage. This type of child may look normal in his features but the behavior patterns and learning ability are deficient. Many times these children are multiply handicapped.

3. Microcephalus. These children are usually very small in sizes. The number and quality of the physical signs vary from case to case and amount of intellectual retardation also varies.

4. Hydrocephalic. These children will vary in degree of retardation. Some may function on a reasonably good level. Others are so severely affected as to be unable to function at all.

5. Phenylketonuria or PKU. PKU is an intolerance of a certain type of protein, causing brain damage. Examination of the urine at birth and special diets can sometimes bring the problem under control and the child will eventually be able to function as a normal child.

Psychological examinations play an important role in recognition of mental deficiency. The mental age (MA) is determined by psychometric test and the intelligence quotient (IQ) is calculated as the ratio of the mental age to the chronological age (CA).

Terminology: IQ—Average 90 to 110. Educable Mentally Retarded (EMR) 50 to 80. (Maximum mental age of 7 to 10 years at maturity.) Trainable Mentally Retarded (TMR) 30 to 50. (Maximum mental age which approaches 3 years or less in adulthood.) The mentally retarded can be trained to the extent that they can function quite normally in social situations. They are also categorized as to social age.

Statistically: 1 mentally retarded child is born every 5 minutes, 126,000 are born every year, 4,200 (1 out of every 30) will be severely retarded and unable to care for themselves. Approximately 12,000 (4 out of 30) will remain intellectually below the 7-year level. 110,000 (the remaining 25 out of 30) are those with mild retardation. Out of an estimated 214 million population, 6.4 million fall below the average intelligence level. Mental retardates and the families they affect, add up to 15 to 20 million people, about 10% of our entire population.

CHARACTERISTICS OF THE MENTALLY RETARDED

They are retarded in terms of physical and mental development.

There is a lack of ability for self-evaluation.

They have difficulty in following a series of directions.

They are unable to perceive relationships and underlying principles readily.

Mental Retardation

There is more frequently a handicap in vision, hearing or motor coordination among the retarded.

Their involvement is with the immediate environment.

They have difficulty in problem solving.

The retardate does not transfer learning as readily as the normal child.

They do better on tasks that involve motor coordination than those involving verbal facility.

They have problems in understanding spatial relationships.

They tend to have shorter attention spans than the normal child.

They tend to come (but not necessarily so) from subculture homes.

They are retarded at maturity.

The academic achievement of the educable mentally retarded child may range from the first- to sixth-grade level.

Frustration proneness is common.

They are often unable to distinguish between acceptable and unacceptable behavior.

They have difficulty in managing social adjustments.

Goals tend to be unrealistic.

They do not readily follow a logical sequence.

The great difference between the normal child and the retarded child is that the former is punished for his "bad" behavior; the latter may not be punished for it, but he is abhorred, which is far worse. It may well be that the tension we see in retarded children is caused not so much by their being prevented from doing things that to them seem perfectly natural as by the horror and revulsion that their inappropriate behavior arouses in all who see it, including, and perhaps above all, their own parents. For, we may be sure that retarded or not, they sense and understand these feelings, which are vastly more effective and terrible than any punishment could possibly be.

Perhaps the answer is to give these children a program in which they use their human powers that will be more interesting than either their fears or the possibility of arousing fear in others. Not that this will be easy to do, but it is where we should aim.[1]

In 1962, Dr. Richard Koch, Director of the Child Development Clinic at Children's Hospital of Los Angeles, reported results of hundreds of youngsters diagnosed as retarded. About one in four actually had normal mental ability. The tragedy is that when such youngsters are treated as if they are retarded, they are in fact likely to become permanently retarded. The error cheats them of normal lives. Some of the children were found to have auditory or visual problems, some were emotionally disturbed. Others had been terribly neglected. "Above all the child should be given, both by parents and professionals, all the love, care and attention that would be lavished on a normal child. To do less is to risk the waste of a useful human life."[2]

Probationer-teacher C. Cleary's first teaching assignment was to a special class of retarded children. Although he had been warned that these children would not accomplish very much, he proceeded to teach them all he could. By the end of the year, many of Cleary's retarded children scored better on standardized achievements tests of reading and arithmetic than did the regular classes.

When Cleary received his dismissal notice he was told that he had grossly neglected the bead stringing, sandbox and other busy-work which were the things that retarded children should do. He had failed to make adequate use of the modeling clay, pegboards and finger paints specially provided by the Excelsior City Special Education Department.[3]

---

[1] John Holt, *"What Parents and Teachers Do to Make Children Fail,"* Pageant, April, 1965, condensed from *How Children Fail* by same author.
[2] "Some Retarded Children Are Not," Reader's Digest, March, 1966.
[3] Lawrence J. Peter and Raymund Hull, *The Peter Principle*, copyright 1969 by William Morrow and Company, Inc.

FATHER DROWNS RETARDED SON

LONDON (UPI)—On a bleak winter's day last month James Price drove his retarded six-year-old son to a secluded river in the English midlands, kissed the boy, then drowned him.

Price, 35, drove straight to the local police station and confessed the deed, saying he did it because his son was "just a living cabbage."

Tuesday, Judge Edward James, in an unusual step in the history of mercy killings in Britain, gave Price his freedom, placing him on probation for a year.

"I am taking a exceptional course in an exceptional case," the judge told Price as he stood weeping in the prisoner's dock in a Worcester courtroom. "I am quite sure that in the passage of time you will be able to forget about this matter."

The judge said he acted as he did partly because of a petition signed by 600 persons in Price's neighborhood in Northfield, near Birmingham, asking the court to show clemency.

During the trial doctors testified that Gordon functioned at the level of a baby aged three to four months, had a short life expectancy, was a spastic and an epileptic.

Price, who pleaded guilty to manslaughter, told the court he decided to drown his son in the River Stour as he was driving the boy back to a hospital for handicapped children after a weekend outing.

"I drove round and round until I came to the river," he said. "I picked Gordon up, knowing full well what I was doing. I kissed him, placed him in the water, and he slowly drifted away in his usual crouched spastic condition.

"I was terribly upset. At last I know my son can rest. To watch him suffer so was a crime in itself."

Chapter 14

Neuromuscular Conditions

The two main neuromuscular conditions that we will discuss here are multiple sclerosis and muscular dystrophy.

MULTIPLE SCLEROSIS may affect the extremities, organs and/or senses, resulting in a loss of function and sensation. The adjustment problems may be due to the awareness that there is no known cure and that the cause is also unknown. The personality traits may range from irritability and depression to that of unusual cheerfulness or a state of euphoria.

The physician may prescribe no specific treatment, however, if the person is able, exercises may aid in the general development of strength and coordination, hopefully progressing to ambulation.

MUSCULAR DYSTROPHY is a disease which causes muscle deterioration. The cause is unknown, however, scientists believe that there may be a faulty metabolism of muscles related to their inability to utilize vitamin E.

The four types of muscular dystrophy include: pseudo-hypertrophic, juvenile, facioscapulohumeral and mixed.

Pseudohypertrophic muscular dystrophy effects more males than females and is hereditary in 35% of the cases. Symptoms appear as bulky calf and forearm muscles with the wasted tissue replaced by fatty deposits and fibrous tissue.

Juvenile muscular dystrophy is hereditary and affects both males and females equally. The symptom evidenced is a more general muscle weakness not limited to the arms and legs, however the progression of muscle deterioration is slower.

Facioscapulohumeral muscular dystrophy effects the face, shoulders and upper arms. As a rule, the person has suffered weakness in the facial muscles since early childhood (being unable to whistle or drink through a straw). Gross symptoms rarely appear before adolescence.

Neuromuscular Conditions

Mixed muscular dystrophy is not inherited and may strike anyone. Weakness usually appears first in the shoulders or pelvic girdle. Progression of deterioration can be either slow or rapid.

As with all handicapped children there is a strong need to be like other children. Mental ability is not affected by muscular dystrophy. The adjustment should be concentrated on adding life to his years rather than years to his life.

The exercise program is mainly concerned with delaying the onset of atrophy in the muscles caused by disuse. The muscles that are still functioning can be aided through movement activities such as walking, building muscular strength, coordination and balance.

Chapter 15

Other Disabilities

POLIOMYELITIS

The characteristic symptoms of poliomyelitis are inflammation of various parts of the central nervous system with particular destruction occurring in the motor cells of the brain and spinal cord. When the impulses to the voluntary muscles become paralyzed, they lose tone, become flaccid and atrophy.

Poliomyelitis is classified as:

1. Abortive (symptoms minor)
2. Nonparalytic type (the central nervous system is involved but the motor cells are not permanently damaged)
3. Paralytic form (either spinal-paralysis of upper and/or lower extremities or trunk muscles; or bulbar—affecting the face, throat and respiratory system)

Adjustments must be made in the students' daily routine to allow more rest and to prevent nervousness. They must do for themselves as much as possible. They must work toward making social adjustments.

Therapeutic exercises serve to reeducate and restore muscle function. An exercise program should progress from passive to assistive, guided, active and resistive. Posture improvement to prevent deformity, should be your primary goal.

The family physician may prescribe exercises to improve strength, coordination and endurance. Care must be taken so that the affected muscle groups are not fatigued or stretched. Protective appliances (braces) must not be removed without the doctor's permission. Balance and flexibility can develop with reeducation exercises.

Other Disabilities

The movement activities should progress from: body mechanics—strengthening weakened muscles of the hip, back and abdomen; to functional training—walking properly, using stairs, and developing ease in changing positions from lying to sitting to standing; followed by rhythmic activities—emphasis on social dancing and square dancing, active games and stunts.

NOTE: Medical guidance must be sought for each new phase of the program.

INGUINAL HERNIA

An inguinal hernia may be congenital or may be caused by underdeveloped abdominal muscles, general muscular disability, injury, operation, or lifting heavy objects.

Under the protection of the teacher, the student must be guided away from activities that may aggravate the condition. Mild exercises can aid in strengthening the abdominal muscles.

Rhythms and games are permissible but the student must be protected from activities where there is danger of a blow to the abdomen. Exercises and activities must be stopped before fatigue is felt and there is danger of increased intra-abdominal pressure. Use of climbing apparatus and activities requiring quick starting and stopping must be restricted.

The following exercises are often prescribed:

1. Lying on back with a pillow under the knees, contract abdominal muscles and exhale, relax and inhale.
2. On back, hands on top of thighs, raise head and shoulders and exhale, relax and inhale with chest up and abdomen in.
3. On back with one hand behind the neck, other hand pressing hernia site; for left hernia, raise head and shoulders twisting to the right and exhale. (Reverse twist for right hernia.) Relax, inhale and repeat.
4. On back with hands at sides, raise head and shoulders part way, slide left hand toward right knee and exhale, relax, inhale and repeat to opposite side.
5. On back, perform bicycle exercise.

6. Sitting with legs straight, feet spread, arms horizontally to the sides, lean forward, touch right toe with left hand and exhale. Return to original position, relax, inhale, and reverse. (Or) touch both toes with alternate hands before returning to original position.

DIABETES MELLITUS

Exercises must be controlled in order to keep the student from overdoing, but any exercise program must be done regularly and daily. The balance of insulin in the body may be upset by too much exercise one day and none the next. Restrict highly organized or competitive sports. Movement activity may stimulate the production of insulin and be an important factor in the amount of insulin needed, however, too much activity may result in insulin shock. The combination of insulin intake and internally secreted insulin through activity may become too excessive in amount. Insulin dosage will have to be adjusted.

PULMONARY TUBERCULOSIS

Pulmonary tuberculosis may be caused by careless coughing, sneezing, spitting, inhalation or by fatigue and poor nutrition. Symptoms include weakness, fatigue, pains in the chest, loss of appetite, chronic coughing, loss of weight or afternoon and night sweats.

Treatment will be a combination of bed rest, medication, and improved nutrition, followed by a gradual physical program. The main problem facing the student recovering from tuberculosis is the prejudice of acquaintances due to misinformation about the disease.

Beneficial exercises must be progressive breathing and chest development, promotion of strength, endurance, rhythms and games. Activity must cease upon signs of fatigue.

ANEMIA, ASTHMA AND ALLERGIES

Anemia is caused by either blood loss (hemorrhage), increased

Other Disabilities

blood destruction, decreased blood production or disturbances in the blood-forming tissues of the body. Lack of desire, strength, endurance, and a tendency to withdraw are characteristic symptoms of the anemic. Social and psychological adjustments are needed. The family physican will prescribe exercises for which little energy is required. The teacher must keep the student from overdoing.

Asthma and allergies are characterized by mental and physical fatigue, emotional upsets, and lack of strength and endurance. Social, psychological and physical adjustments are needed.

The family physician will prescribe exercises to increase lung capacity and to aid in reducing the number and severity of attacks. The asthmatic must be placed in a situation where he is in competition with himself rather than with others.

MALNUTRITION

For the obese child the family physician will prescribe, along with an improved diet, an exercise program which serves to develop speed, agility and endurance as well as activities which promote social adjustment. The underweight child will have the same needs in addition to development of strength. The exercise program designed for these children will be influenced by the particular needs of the individual with emphasis on conditioning and development.

SKIN PROBLEMS

Numerous types of skin problems include acne, boils, carbuncles, moles, scabies, impetigo contagioso, eczema, and ringworm. The teacher must maintain control of any contagious skin problem and restrict the student from contact activities and vigorous activities. Social adjustment is most difficult but most important.

Part IV

KINEMATICS

Chapter 16

The Classroom

THE KINEMATICS METHOD

Each Kinematics class should include the following:

I. Stationary exercises
   A. Across-the-floor movements stressing large muscle development, strength, balance, flexibility, coordination and rhythm (several repetitions of each).
      1. Bend forward, backward and side to side.
      2. Bend knees, straighten, rise to balls of feet.
      3. Stretch alternate legs to front, side and back.
      4. Sit, legs stretched, bounce body forward and up.
   B. Educational activities
      1. Naming body parts, moving each part or touching.
      2. Naming and moving in various directions.
      3. Move each body part in different directions.
      4. Move at high, low and medium levels.
      5. Move in a small space and in a large space.

II. Locomotor exercises
   A. Across-the-floor movements stressing large muscle development, agility, speed and endurance.
      1. Walk
      2. Run
      3. March
      4. Slide
      5. Jump
      6. Hop
      7. Skip
      8. Turn

B. Circle exercises, repeat 1 through 8 clapping rhythms.
III. Therapeutic exercises for each student's individual needs.
IV. Rhythmic dance patterns or creative movement.

NOTE: Remember that each stretching movement must be followed by a relaxing movement. Each forward movement by a backward movement. And each movement to the right followed by a movement to the left. The exception would be in a case where one side is overdeveloped; concentrate on the underdeveloped side.

EQUIPMENT (basic needs):
  Phonograph or tape recorder
  Set of records (see record index)

EQUIPMENT (additional suggestions):
  Balls (both large and small); yarn balls are easier to catch
  Beanbags
  Hula Hoops
  Ropes
  Stretch ropes
  Cardboard discs (for stepping-stones in balance exercises), carpet scraps with rubber backing are much safer.
  Mats (use folded blankets or old covered mattresses)
  Rhythm instruments
    Maracas
    Tambourines
    Small fruit juice cans with dried beans inside
    Large coffee cans for drums
    Wrist bells and ankle bells
    Bell-sash belt
    Finger cymbals

EDUCATIONAL EQUIPMENT:
  Large numbers (could be drawn on cardboard discs above)
  Large letters (can be cut out of textured material for the blind)
  Large sketches or poster pictures for visual motivation
  Large imitation clock

The Classroom

Large model of a traffic signal
Chalk board and chalk

LESSON PLANS

Lesson plans should be designed to develop the complete child; physical, psychological, intellectual, social and emotional. Movement skills must include agility, strength, flexibility, speed, balance, endurance and coordination. Rhythmic exercises in which each child follows his own inner rhythm develop self-awareness. Fast reaction, the ability to initiate movement, change direction or adjust position all contribute to agility. Balance exercises must include three kinds of balance: static (not moving), dynamic (moving), and object (supporting an object). Endurance exercises are not recommended for children under the age of eight.

Your students will be more secure with a structured class in the beginning (you do all of the planning); then later, when they have acquired a sufficient vocabulary of movements, try a more creative and experimental class (let them do some of the planning). Transfer of decision making from the teacher to the student should be gradually introduced according to the child's ability to accept this responsibility.

PRINCIPLES OF LEARNING

1. Practice sessions should include rest periods. Hold interest by changing the activity if you notice boredom.
2. Twenty minutes a day may be enough for some skills.
3. Begin with simple tasks, add a part and a third part until the whole complex skill is assimilated. (Serial Memory Ability)
4. Do not "overteach," allow self-direction.
5. Up to a point, retention is increased by overlearning.
6. The student should understand in advance that he is expected to remember the activity and the reason for the activity. (Pattern Recognition)
7. Rhythmic movements are easier to remember than unrelated movements.

CAUTION: The teacher who tries to meet all the children's needs is bound to exhaust herself beyond her ability—into realms of social work, psychotherapy, and even medicine. Not only is she inflicting harm, but she is ignoring the job she is hired to do—that of promoting learning. The teacher may complain that she cannot depend on other people to solve the children's problems. But can she be depended upon to solve the children's learning problems?

MODIFYING CLASSROOM BEHAVIOR

1. The student must succeed. With or without your help, he must complete the task given him.
2. If each step the student takes is small, he is not likely to fail.
3. Success and good behavior must be reinforced by praise.
4. If no one pays attention to bad behavior or verbal outbursts this kind of behavior will likely stop.
5. If ignoring fails, the student exhibiting bad behavior should be isolated. Do not warn or threaten.
6. A well-mannered child from a familiar story may serve as a model for good student behavior.
7. Positive reinforcement for good behavior must be consistent. The teacher must acknowledge it each time it occurs.
8. Once the behavior is well established, reinforcement and rewarding should be intermittent.

PUNISHMENT

1. When punished the child may strike back at the object or a socially irrelevant object (e.g., throw a book).
2. The teacher may come to be identified as an enemy of the child especially if punishment is given in an angry, threatening manner.
3. The effects of harsh punishment have been shown to last only for a short time.
4. The bad behavior may be suppressed only in the presence of the teacher.

DISCIPLINE

Children learn control while waiting in line, in performing movement sequences accurately, in trying to stop their movements at predetermined points, in staying inside a particular space without disturbing the movement of others, and in moving with partners or in unison with a group. Clinical experience proves that control of movement leads to improved control of behavior and self-control. This can also be achieved by alternating quiet exercises with exhilarating ones. It is up to the teacher to know when students need to be "turned on" and when they need to be "turned off" and to have the appropriate music available with which to bring this about.

CONTINGENCY CONTRACTING OR "GRANDMA'S RULE"

1. Contingency contracting is simply stating to your students that "Something pleasant will happen if . . ." As in any contract, benefits are forthcoming only when you live up to the terms of the agreement.
2. The rule is that the contract must be fair, clear, honest, positive and systematic.
3. The contract should say: "If you accomplish such and such, you will be rewarded with such and such," not, "If you do what I tell you to do, I will reward you." Or, "If you do not do what I tell you to do, I will not reward you." Or "I was going to reward you, but now that you've been bad, I will not."
4. Reinforcers, rewards or praise must come immediately after the accomplishment.
5. A gradual transition should be made from teacher-controlled contracting to student-controlled contracting.
    a. The teacher determines both the task and the reward.
    b. The teacher and student jointly determine the task, the teacher determines the reward, or vice-versa.

      c. As in *b*, but the teacher still has control over the extent of the task and the amount of reward.

      d. The student now has full control over either task or reward and the teacher and student jointly determine the other.

      e. Student determines both task and reward.

6. Make sure the student is able to keep the agreement by keeping the task within his ability to succeed.

EDUCATIONAL ENGINEERING

Step 1. *Attention.* You can gain the attention of your students by verbal direction, teaching on a one-to-one basis, wearing bright-colored clothing, using appropriate music and dancing in a circle. You must have their attention before proceeding to the next step.

Step 2. *Response.* If your students are not responding by actively participating you may need to lower your level of expectation, guarantee his success, allow the student to withdraw for a short time, or motivate social contact, approval, and attention from the others in the class. If the problem is boredom, you may need to raise your level of expectation with challenging activities.

Step 3. *Order.* This level deals with helping the child to adapt to class routine, follow directions, complete techniques and combinations and control behavior and movements. The class should have:

      a. A well-defined beginning, e.g., a greeting to the teacher, lining up ready to begin, and end including a systematic series of exercises and movements leading to the completion of a routine.

      b. A gradual increase in difficult steps and movements.

      c. Total involvement in a satisfying activity by each student.

Step 4. *Exploratory.* The student is now ready to initiate, create and interpret movements into a dance form of his own. The teacher must be sensitive enough to know when the students are ready for this level and not anticipate it too soon.

The Classroom

Step 5. *Social.* The student maintains appropriate behavior in class, communicates well with the instructor and peers and displays the ability to adapt to a variety of situations.

Step 6. *Mastery.* This level deals with the student's ability to function independently in self-care, concept formation, and problem solving within his intellectual and physical potentials.

Step 7. *Achievement.* This level may include the student's ability to initiate and create a routine and rehearse it until it is ready to be performed before an audience, e.g., parents and teachers. Teacher control needed should be minimal at this step.

PROBLEM SOLVING

Many times, movement exploration may lead directly into a creative dance experience. It is the participation itself, however, rather than any polished performance, that matters. To solve any problem, a child needs to think about it, to discuss it, and then to carry it out through experimental activities.

1. How can you gallop? What animals gallop? How does it look? How does it feel? What body parts make us gallop? Can you gallop in different directions? How high can you gallop? Can three people gallop together?
2. Try skipping around the room without touching anyone else. Try to skip with a partner. Can you skip backwards or sideways? (Continue using other locomotor activities.)
3. Try to run and cover as much space as you can, without touching anyone else, before the music stops.
4. Find your own space on the floor where you won't touch anyone else. Reach out as wide as you can. How high is your space? What is the smallest shape you can make? The largest? Can you make the first letter of your name with your body? A circle? A triangle?
5. What have you seen that swings? Can you swing one part of your body? Another part? Two parts at the same time? Can you swing with a partner?

6. Can you move your hand in a slow, smooth, steady way, like the moon coming up, or a flower growing?
7. Can you move very fast without leaving your space? Can you move one part of your body fast and another slow? Can you move a part of your body in a jerky way?
8. Move as though you were very heavy, like a giant or an elephant or a truck going uphill. How would you lift something heavy? Push something heavy? Pull something?
9. Can you move as if you were very light? What is light? (feathers, balloons, birds) Can you fall down lightly without making any noise?
10. Move your head four times, your arms four times, your legs, now all of you.
11. How do you move when you're happy? Sad? Tired? Scared?

SPACE

I have saved "space" for the last of this chapter because it is usually the last thing people consider when arranging for movement classes. If you are making the arrangements and you must ask permission for the use of a room, get the measurements. Don't settle for, "Oh, there's plenty of room." This usually means you have six square feet for a class of twelve students.

Children need space to move, to feel free, to explore. You must have a room that is large enough to allow for vigorous movement without any obstacles. A multipurpose room, auditorium, gymnasium, or dance studio is most desirable. (NOTE: Children who are emotionally disturbed may tend to feel less secure in a spacious auditorium.) In a very large room or outside, you can mark off boundaries with chalk or white shoe polish or ropes.

If a room of this type is positively unavailable, the classroom can be cleared in a short time by pushing chairs, tables and desks to the outer edges. This can be done very efficiently if routine is established and the children practice a few times. If the desks are permanently fixed to the floor, then plan your lessons with movements that can be done between them, under them, over them, and around them, but use all the space you have.

Jackets and sweaters should be removed so the students can be as free as possible. If the room is cool, which it must be, sweaters should be kept on until the warm-up exercises. Leather shoes and stocking feet are slippery, so the students should have tennis shoes or be barefooted. This will avoid many accidents.

A wooden floor is best, since it will be more resilient and will be easier on the wear and tear of the body when jumping and running. Although the spine, knees and feet, if used properly, will act as shock absorbers, the amount of jumping during classtime must be lessened if you are working on a cement floor.

The weather also plays an important role in the type of exercises included in your lesson plans. In hot weather the muscles are more supple and stretching movements are easier. They must be done all year around, but go easy in the wintertime. As you know, breathing can be difficult when the weather is hot and humid, so save the vigorous activities for cold weather. This does not mean to eliminate completely the strenuous exercises during the summer, but use your common sense and don't overdo. Damp rainy weather also plays tricks on our bodies. Rainy days are good days to teach relaxing exercises and creative movement.

Chapter 17

Passive, Range of Motion Exercises

For those persons confined to a bed or wheelchair or having activity restricted in any way by the family physician, permission to proceed with the following exercises must be given by the doctor.

These exercises should be done daily, each motion ten times. They should be given slowly and carefully. It is advisable that a warm bath be given before the exercises to relax the muscles.

UPPER EXTREMITIES

Lying on back:

1. Holding the arm at the elbow and wrist, take it straight forward and upward toward the ear.
2. Holding the arm at elbow and wrist, take it straight out from side and upward toward the ear.
3. Take the arm out to the side away from the body; with the elbow bent, grasp the arm at the elbow and wrist and bend it forward and down toward the bed, and backward toward the bed.
4. Grasping the arm at the elbow and wrist, and, with the palm of the hand facing the patient, bend and straighten the elbow.
5. With the elbow bent to a right angle, grasp the arm above the wrist, and turn the forearm until the palm of the hand faces the patient; then turn the palm away from the patient.
6. Bend the wrist forward and backward and a little to each side.
7. Take the thumb straight out to the side from the hand until it forms a right angle with the hand.

8. Take the thumb straight forward from the palm of the hand until it forms a right angle with the hand.
9. Take the thumb across in front of the palm of the hand until it touches the tip of the little finger.
10. Spread fingers apart and together again.
11. Keeping wrist straight, bend and straighten each joint of each finger; then bend fingers together to make a fist and straighten them.

LOWER EXTREMITIES

Lying on back:

1. Raise the leg allowing the knee to bend, and then gently force the knee toward the chest while holding the other leg down.
2. Raise the leg straight from the bed, with one hand over the kneecap to keep it straight and the other hand under the heel until it is at a right angle to the hips. Remember to keep the hips and the opposite leg flat. This exercise may also be done by bending the knee and then straightening the leg.
3. With one leg flat on the bed, bring the other leg toward the chest and place the foot flat on the bed. Bend the leg that is up in toward the leg that is on the bed, and then out and down toward the bed. This exercise may also be done by bending the hip and the knee to right angles and grasping the leg at knee and ankle. Rotate the foot and lower leg in toward the leg that is on the bed; then carefully out away from the leg.
4. Take the leg straight out to the side. Keep the knee flat by holding the leg in the crook of your arm and placing your hand over the kneecap. Let the opposite leg lie flat on the bed in a normal position.
5. With the leg straight, grasp the heel and rest your forearm on the ball of the patient's foot. Then pull the upper part of the foot back toward the knee until it is just a fraction past 90 degrees from the bed.

6. Bring the forepart of the foot up and turn it in.
7. Bring the forepart of the foot up and turn it out.

NOTE: These exercises must be done carefully and should be demonstrated by a professional therapist at the first session. Gradually the patient should begin, if possible, performing the exercises with less help, eventually performing them alone.

ACTIVE ASSISTIVE

Lying on back:

1. For relaxing, pretend that your body is very heavy, so heavy that no one can lift you. Feel yourself sinking into the bed.
2. Concentrate on tightening and relaxing each part of your body separately beginning with the toes, then the feet, ankles, legs, hips, stomach, back, etc., until you have reached the top of your head.
3. Pretend that each part of your body is a balloon. Mentally blow up each balloon and then slowly let out all of the air.
4. Pretend you are a puppet and someone is pulling the strings connected to each part of your body. Try to raise each leg and each arm as if the strings are being pulled, and then let go and drop each limb onto the bed.
5. Raise the upper body onto left elbow, reach right hand toward left leg. Ease right hand under thigh and pull leg toward chest. Still holding left thigh, return to the back-lying position and use both hands, to pull knee toward chest.
6. Hold thigh with one hand and grasp ankle with the other hand. Use your hand to wriggle your toes, move your foot right and left and in a circle. You may use both hands to manipulate movement of the foot.
7. Hold thigh again with both hands and press leg across the body to the right and up again and down toward the bed.

NOTE: If at any time during the above exercises the patient begins to feel fatigued, stop and repeat one of the relaxing exercises before continuing. Repeat the complete combination of move-

Passive, Range of Motion Exercises

ments in exercises 5 through 7 on the opposite side.

To assist the patient in learning to turn over by himself for the next set of exercises there must be rails on each side of the bed or an assistant on each side.

Reach left hand under right hip, pulling leg across body to left. Scoot left hip to right (with assistance) and using right hand to pull against assistant, bed rail or side of bed, turn over onto stomach.

8. Pushing with both hands, raise body as high as possible, eventually progressing to the point where both arms are straight.
9. Lift hips and push hips back until you are resting on your hands and knees. Arch the back and round the back, alternating movements.
10. Push hips back further so that you are sitting on your heels. Starting at the top of your head, roll down, rounding the back until the top of your head touches the bed. Reverse movement, starting at the tailbone and roll up to sitting position.
11. When patient is able to perform exercises in a sitting position head and shoulder exercises can be added to the session.
12. Finger, hand, wrist and arm exercises should be used during each session.
13. For strengthening the back, the following exercise is recommended. Lying on stomach, bend knees, bringing feet toward the hips. Reach right hand back to grasp right ankle, left hand to left ankle and pull head and shoulders toward feet. Relax and repeat.
14. For children, to develop body awareness with tactile stimulation, have the child move each body part as it is touched and call out the names of each part; toe, knee, nose, hip, stomach, lower leg, cheek, thigh, neck, knee, chest, cheek, toe, foot, arm, fingers, forehead, ankle, etc.

NOTE: Always use background music with a relaxing rhythm.

Chapter 18

Stationary Exercises

EXPLANATION OF TERMS

1. *Gross Motor development:* development of the large muscles of the body contributing to strength, good posture, coordination and balance.
2. *Fine motor development:* development of the small muscles contributing to eye-hand coordination, eye-foot coordination and ability to isolate movement in various parts of the body.
3. *Perceptual or sensory-motor development:* conscious movement and the ability to initiate and direct movement by auditory, visual, tactile and kinesthetic stimuli. Knowledge and concepts of big and little, on, under, around, over, inside and outside.
4. *Self-awareness, self-concept or self-image:* a feeling of identity necessary for the development of creativity and individual accomplishment and achievement.
5. *Social-awareness:* being aware of others and having the ability to work and play with others toward a common goal.
6. *Time and space awareness:* knowledge of the environment, of past and future, of space beyond what can be seen and touched. Knowledge of slow and fast, even and uneven rhythms.
7. *Personal Space:* the sphere surrounding a person that he can touch with outstretched limbs, while remaining stationary.
8. *Common or social space:* the area used by the group, where each moves freely.
9. *Laterality:* preference for, or superiority of, one side of the body, e.g., the right-dominate person prefers to use the right hand, right eye and right foot, in writing, reading and motor activities. Lateral dominance is not always related to learning.
10. *Directionality:* knowledge of left and right. Directionality does have an effect on learning abilty. The child has difficulty in

orienting himself in reading, writing letters and numbers in the correct direction.

11. *Midline:* the imaginary vertical line in the center of the body. The child with a midline problem is unable to swing his right arm to the left without turning his body. The same is true of the footwork. In reading, the child reads the left side of the page with his left eye until he reaches the midline and the right eye takes over. In normal children, both eyes work together.

12. *Earthbound:* the inability to perform motor activities that demand both feet to leave the floor simultaneously, such as skipping, hopping, jumping, leaping. This is most prevalent in blind and retarded children without training.

13. *Hyperactive (Hyperkinetic):* continuously in motion, jerky, rapid, abrupt, interfering with the completion of their own tasks and the tasks of others, not goal-directed.

14. *Hypoactive (Hypokinetic):* moving slowly, sitting motionless for long periods, low energy level, needing more time before reacting to directions.

15. *Serial memory ability:* remembering a sequence of activities previously learned.

16. *Pattern recognition:* being able to recognize a step or movement performed by a fellow student, a floor pattern (circle, square, line) previously learned, and being able to copy the teacher's movements.

The following exercises are those which can be done in a small space and may be used to initiate a short five minute exercise break between academic lessons:

1. Touching toes.
2. Rising to the balls of the feet and down.
3. Bending knees, making certain that the heels are down.
4. Head movements, up and down, side to side and circle.
5. Finger and hand exercises. Crumpling paper, stretching rubber bands.
6. Jumping in place.
7. Arm and shoulder exercises.

8. Running in place, marching in place.
9. Body bends, backward and forward, side to side, and circle.
10. Naming and touching body parts and naming directions.
11. Clapping names in rhythm, snapping fingers, stamping.
12. Balance exercises.
13. Deep-breathing exercises.
14. Creativity: let the students make up their own movements in their own personal space by imitating an animal, toy, or simply responding to the rhythm of the music.

NOTE: Items 1 to 5 can be used for gradual warm-up and to promote flexibility.

Items 2 and 3 prepare students for properly executed jumps—begin with knee bend, knees directly over toes, heels down, hips directly above and between feet, stomach pulled in, back straight and neck relaxed. For push-off, knees straighten, heels leave floor first, followed by balls of the feet and toes, back remains straight. For landing, toes return to the floor first, followed by balls of feet and heels, knees bend over toes, heels must not be allowed to bounce off of floor, back remains straight throughout. Proper jumping technique promotes higher, more gracefully controlled jumps and eliminates shock to the vertebrae, knee joints and feet.

Items 6 to 9 give an example of how to alternate strenuous activities with those requiring less energy, thus preventing fatigue or hyperexcitability.

Items 10 to 13 promote gradual relaxation of the mind and body in preparing for 14—creativity.

The following stationary exercises are those which can be done in a fairly small space. Classroom desks may be moved against the wall leaving the center area clear for movement.

1. Stationary crawling: e.g., be an alligator or a snake, or swim.
2. Angels in the Snow: students must be able to move each arm and leg separately, alternate arm and leg separately, arm and leg on same side, and both arms and legs, on command.

Stationary Exercises

3. Stomach-rock or rocking-horse exercise.
4. Push-ups: raising shoulders and upper body only; raising shoulders, upper body and hips (weight on hand and knees); or if student is able, do regular push-ups. If unable to perform regular push-ups, gradual progression should be made toward this goal to promote strength and endurance.
5. Sit-ups, with knees bent.
6. Leg-lifts: lift one leg (knee bent, then straighten) repeat with opposite leg and progress to lifting both legs straight and together.
7. Sitting stretches: sit with legs together or apart and attempt to touch hands to feet keeping knees straight, progress to touching head to knee or to floor.
8. Bicycle exercise: can be done with student either sitting in a chair, lying on back or with weight on shoulders, upper back and hips lifted as high as possible.
9. Rolling a ball: student may be in sitting, standing or crawling position and use hands, feet or head to push the ball.
10. Leg-swings and arm-swings for balance and coordination.

In chapter 20 you will learn more about dance composition, but, even with this limited vocabulary of movement, you can begin to experiment by combining any of the previous movements to create a "dance."

EXAMPLE ONE: For teaching body parts.
"Boa Constrictor"
  Record: Peter, Paul and Mary *(Peter, Paul and Mommy)*
    Warner Brothers Records LP #1785
"I'm being swallowed by a boa constrictor" (repeated three times).
  Movement: Wriggle down and up imitating a snake.
"And I don't like it very much. Oh no, oh no, he swallowed my toe, he swallowed my toe."
  Movement: Bounce body forward, touching toes four times . . .
"Oh gee, oh gee, he's up to my knee, he's up to my knee."

Movement: Bounce body forward, touching knees four times.
"Oh fiddle, oh fiddle, he's reached my middle, he's reached my middle."
Movement: Hold waist and bend side to side; repeat.
"Oh heck, oh heck, he's up to my neck, he's up to my neck."
Movement: Hold neck and bend head side to side; repeat.
"Oh dread, oh dread, he swallowed my . . ."
Movement: Hold head and move it side to side.
"Oops!"
Movement: Quickly bend knees and stoop as low as possible, holding hands over head, head tucked down.

EXAMPLE TWO: For teaching coordination, balance and flexibility.
"Swimmy Swim Swim"
Record: Woody Guthrie *(Children's Songs)* Golden Records LP # 238
"I like to swim in my water, like to swim in my water, water water and water, swim swim swimmy oh swim," etc.
> *Movement:* Stationary crawling, beginning with arm and leg on same side, progressing to alternate arm and leg. For those unable to do this, sit on floor or chair and imitate swimming movements with arms.

"I like to kick in my water," etc.
> *Movement:* Bicycle exercise.

"I like to splash in my water," etc.
> *Movement:* Rocking-horse exercise, slapping hands on the floor with each rocking movement. Those unable to do this may simply perform the splashing movements with hands.

"I like to float in my water," etc.
> *Movement:* Angels in the Snow exercise. Continue to end of song adding arm movements and kicking movements while balancing on hips.

Teaching suggestions: The student must learn each movement separately before combining them. Next, combine two movements until they can be performed easily, adding the third and fourth gradually.

Other Stationary Exercises

BODY AWARENESS

Isometric
1. Interlace fingers of both hands. Press palms hard together. Hold while teacher counts to three, relax pressure. Repeat.
2. Bend head to one side, keeping face forward so that maximum tension is felt in the neck muscles of the opposite side. Keep other body parts still. Hold while teacher counts to three; relax; return to original position. Bend to other side. (I suggest that increased tension and awareness will be felt if the student pushes his hand against his head to offer resistance.)
3. Sit cross-legged on the floor, hands behind neck, fingers interlaced. Press hands against neck and, rounding back, bend forward as far as possible, pressing neck down with hands. Hold while teacher counts to three. Repeat; lie on floor; relax.
4. Same as above except bend head backward with hands pressing against forehead, try to raise head to erect position to a count of three; relax; repeat.
5. Lie on back on the floor. Knees may be bent slightly. Stretch arms and hands forward; raise head, neck and shoulders from floor. Hold to count of three. Relax.
6. Lie on back on the floor, knees bent, feet flat. Contract abdomen. Hold and breathe in and out deeply three times. Relax.
7. Sit on the floor, knees bent, feet on floor. Shift weight backward onto lower back, letting feet rise from floor. At the same time bend head and shoulders forward. Hold while teacher counts to three. Return to sitting position without using hands if possible.

Relaxation
8. Try to "melt" into the floor while on back either with eyes closed or open. (Pretend to be ice cream.)
9. Sit on the floor, and pretend to be a rag doll. Done with eyes open or closed.

10. Practice falling like rag dolls. This can only be done on a mat and after the students have been shown how a rag doll falls.
11. While "melting" into the floor or during any rest period, become aware of the rhythmic activity of the heartbeat and breathing either with eyes open or closed.
12. Head roll. Lie on the floor and roll head to right, look right, roll head to left, look left. Can also be done with eyes closed.
13. Shoulder roll. Same as above only roll shoulders as well as head.
14. Sit with legs crossed and let the body droop forward in a relaxed position. Slowly rotate head, shoulders, and trunk, but do not shift weight so much that balance is lost.
15. The students lie on their backs, relaxed. They slowly raise their knees, letting the feet slide along the floor toward the trunk. They then relax the tension in the leg muscles and let the feet slide back to the original position.
16. As above, but when the feet are near the trunk, let the knees and lower legs fall to the right so that they touch the floor. Then swing them with as little effort as possible to the left, and so on, from side to side in an easy motion, keeping the trunk relaxed on the floor.
17. The students lie on their backs, relaxed. They slowly raise their arms to a vertical position; then they relax the muscles and let the arms drop to the floor.
18. Hold on the rail, chair or other fixed object and let one leg swing easily and freely. Forward and back, to the side, or in a circle. Progress to balancing without support.
19. Relax the arm and leg muscles by shaking them. Stand with the weight on one foot and shake the arms and the free leg, or sit or lie down and shake arms and legs.

Tactile stimulation with resistance

20. Press against the floor or other rigid surface with the whole body in both face up and face down positions, with eyes open or closed.
21. Kneel with hands on floor below shoulders and slightly

Stationary Exercises 103

turned inward, thighs at right angles to the floor, back straight. Try to press "holes" in the floor with hands. Count to three. Relax.

Posture

22. Stand with back about four inches from the wall. Lean backward and press against the wall with back as hard as possible. Still pressing, slide down to a position as if sitting on a chair, with knees bent and back flat against the wall. Slide up to original position. Repeat.
23. The students stand about ten inches from the wall, facing it. They lean toward it and pretend to push it back with their hands.

BALANCE

Static

1. *To stand on tiptoes:* For all tiptoe positions, the heels should not be raised as far as possible, only as far as comfortable. Raise heels, hold three to five counts and lower heels. Better results will be achieved if the student holds a large ball with both hands high overhead.
2. *For coordination:* Sit on the floor with legs straight and apart, arms stretched to side level with shoulders. Bend slowly and rhythmically to the right and left, keeping arms and back straight. May also be done standing with feet apart or standing on one leg.
3. *For flexibility:* Stand on one foot and bend forward, raising the free foot backward until the trunk and free leg are parallel to the floor, arms held to the sides. Hold to count of three, return to standing position and reverse. Maye also be done with leg raised to the front or the students may be asked to balance on one leg in any position they desire.
4. Balance on one leg and swing free leg forward and backward, alternating feet. May also be done swinging leg to front, side and back. (To avoid swayback, the backward swing should not be emphasized or the trunk held rigid.)
5. *For strength in abdominal and back muscles:* Sit on the floor,

legs together and straight, arms out to side, level with shoulders. Raise legs as high as possible, keeping them straight, maintaining balance on buttocks without using hand support.

6. *For strength in abdominal muscles and back and for flexibility:* Students assume same position as in bicycle exercises but move legs back and forth like scissors, keeping legs straight. May also reach back until feet touch the floor behind head.
7. *Static, dynamic, coordination and agility:* Students are paired off and one adopts various balancing positions or poses that the other imitates. Reverse the roles.
8. *Static:* In crawling position, students practice raising one arm, one leg, then arm and leg on same side, then arm and leg on alternate sides. Repeat with eyes closed.

FLEXIBILITY

1. *Shoulder joints and strength in shoulder muscles:* Sit with arms stretched out to sides, backs straight, and make small circular movements backwards with arms, gradually increasing the size of the circles. May be done standing, using various tempos, making circles in front of the body or above the head. Students may be asked to invent their own circular movements in any position they desire.
2. *Spine and leg muscles:* Sit on the floor, legs spread and straight. Lean forward, grasp one ankle with both hands and pull head down toward leg. Repeat, alternating sides. May grasp calf of leg if unable to reach ankle.
3. *Spine, leg muscles, coordination, crossing the midline:* Sit on the floor, legs straight and as far apart as possible. Slide a beanbag forward and back to the thigh of each leg alternately, bending forward as far as possible while keeping the legs straight. May be done holding the beanbag high above the head and bending forward to touch the foot with the beanbag. May also be done in standing position.

Stationary Exercises

4. *Spine, knee joints, static balance:* Sit on the floor holding one foot with both hands and try to touch the foot to the head. May also be done in standing position.
5. *Hip, knee joints:* Lie on back and raise one leg as high as possible, keeping the leg straight and the other leg flat on the floor. Lower the leg slowly and repeat with other leg. May be done with student using both hands to pull the leg gradually closer to the chest.
6. Stand with feet together and slowly raise one leg forward and as high as possible, then slowly lower it, keeping both legs and back straight. Repeat with alternate leg. Repeat to the side and to the back. (To avoid swayback, the backward lift should not be emphasized nor the trunk held rigid.)
7. *Spine, static balance:* Sit on the floor with legs straight in front. Place hands on hips and turn as far as possible to the right without moving legs or hips and keeping back straight. Repeat to left and alternate in smooth rhythmic motion. May be done in standing position and with hands stretched above head.
8. *Spine, leg muscles:* In standing position, feet apart, bend forward and reach back between legs making a mark on the floor with chalk. May also be done with clip type clothespin. Reach back between legs and try to clip clothespin to dress or shirt.
9. *Spine, relaxation:* Sit cross-legged and rotate head clockwise then counterclockwise, very slowly for maximum stretch, keeping trunk and shoulders still. May progress to rotating upper body and shoulders.
10. *Spine, coordination, agility, static balance:* Students stand in single file, legs apart. The front child passes a large ball or beanbag between his legs to the child behind, who passes the ball backward over his head. As the child in back receives the ball he brings it to the front and the process is repeated. Children should be far apart to require stretch.
11. *Spine, agility, strength in leg and abdominal muscles:* Sit on the floor, knees close to the chest, arms wrapped around

knees, head bent forward, feet off floor. Roll backward and forward. Progress to swinging up to a standing position by thrusting arms forward.

12. *Spine, agility:* Lie on floor on back. Raise legs overhead with knees bent and bring knees down on either side of the head as close to the floor as possible. Roll backward to a sitting position. May progress from using hands to push against floor to rolling without using hands.
13. *Hip and knee joints, static balance:* Stand straight and lace fingers together making a ring with arms. Pull one knee up to chest, keeping the supporting leg and back as straight as possible, although head and neck may bend forward. Maintain position for count of three. Progress to stepping through the ring, first with one leg and then the other.
14. *Spine:* Assume crawling position on hands and knees, back straight. Round back, pulling head down, and follow by bringing head up and arching back.

COMBINATION

1. *Crossing the midline, flexibility of shoulder, hip and spine and static balance:* The teacher should demonstrate with pictures and objects to show what a pully is. Lie on back and raise each leg alternately to a vertical position, keeping legs straight. Raise each arm alternately, keeping legs flat on floor. Raise both arms and both legs alternately. Alternately raise left leg and left arm with right leg and right arm. Progress to standing position raising legs alternately and touching them with opposite hand. Children with difficulty in balancing may stand against a wall for support.
2. Two students stand about five feet apart and throw a beanbag or ball to each other. Gradually increase the distance. Clap hands before catching it. Jump before catching it.
3. Students sit on floor in circle of eight to ten feet in diameter, legs straight and apart. One child sits in center, and rolls ball to a child in the circle. That child rolls ball back to center child, who continues to the next child in the circle. When

each child has had a turn, another child goes to the center. Variation: do not let students use hands to roll the ball, use feet instead.

AGILITY

1. *Strength in abdominal muscles:* Lie on one side, arms and legs stretched forward at right angles to the body. Roll to the other side. Repeat with arms and legs bent. Progress to three rolls, gradually gaining momentum and rise quickly to a standing position on the third swing.

Chapter 19

Locomotor Exercises

EXPLANATION OF TERMS

1. Sliding can be either:
   a. moving feet forward, backward or to the side without leaving the floor as if on ice or through snow. In ballet this is termed *glissade* or gliding.
   b. moving feet forward, backward or to the side with one foot always leading. In ballet this is termed *chasse* or chasing.
2. *Prancing* is similar to running except the feet are raised to the front, toes pointed.
3. *Galloping* can be either sliding as in (b) above or prancing, depending on whether your background education has been in ballet training or folk dancing.
4. *Jumping* is taking off from two feet and landing on two feet.
5. *Hopping* is taking off from one foot and landing on the same foot.
6. *Leaping* is taking off from one foot and landing on the other.

How can you move from one side of the room to the other? Demonstrate type of rhythm to be used for each by clapping, using drum or rhythm instrument.

1. Walking (forward, backward, low level, high level, diagonal, turning or using crossover steps). Arms should swing alternately.
2. Running (same as walking)
3. Skipping (same as walking)
4. Galloping
5. Prancing

Locomotor Exercises

6. Jumping (forward, backward, sideways)
7. Hopping
8. Leaping
9. Sliding
10. Marching
11. Crawling (forward, backward, sideways)
12. Turning
13. Rolling
14. Crab walk (forward, backward, sideways)
15. Somersaults (backward and forward)
16. Duck walk

Suggested activities: Use any of the above activities in teams or groups. Experiment with various formations, e.g., straight line, circle, diagonal line; or patterns and directions, e.g., moving about the room scattered, moving in a square, or moving in and out of a circle. Bind children will feel more secure if holding hands in circle.

Warning: Younger children (ages 3 to 4) may have difficulty in activities requiring a circle formation. They tend to mirror the movements of those standing opposite from them. Even some adults have difficulty in circle formations.

SUGGESTED LOCOMOTOR EXERCISES

BODY AWARENESS

1. *Tactile stimulation with resilient resistance:* Two students stand back to back. One pushes backward; the other provides resistance but only to the degree that he is pushed slowly across the room. Reverse the roles. Variation: Stand facing each other, palms touching. Variation: Stand facing each other with foreheads touching.

2. *Tactile and kinesthetic stimulation:* Crawl through a tunnel, roll on the floor or do somersaults using weighted armbands or ankle bands or with some weighted objected (beanbag) attached to a part of their bodies or carried on their heads or in their hands.

3. *Avoiding collision:* Form an obstacle course and direct the students to: e.g., crawl under a table, step over a chair, jump over lines drawn on the floor, go around a chair or other children, etc. *Variation:* The students are directed to run the obstacle course in different directions or from different starting positions or move within a designated area, avoiding collisions.

4. *Right and left discrimination:* Practice making quarter-turns, half-turns, and full-turns to right or left using the following method: Face teacher, point to teacher, carry arm to right pointing right, turn to face right arm, continue until making one complete turn. Say the action words, "point right," "turn right." *Variation:* Divide the class into those who are left-handed and those who are right-handed. Left-handed children turn left only, right-handed turn right only. *Variation:* Perform turns by jumping and call out action words "Quarter-turn right," "quarter-turn left," progressing to "whole turn."

5. *Right and left discrimination:* Beanbags or carpet squares are placed on the floor to form a course. Each student, in turn, proceeds from the starting place to the goal by jumping over the objects. After jumping over the first, the student should turn toward the second calling out in which direction he has turned. He continues until he has completed the course. The course should be arranged so that the students must make both left and right turns.

6. *Right and left discrimination, perception of spatial relationships:* Each child, in turn, follows a chalk line until he encounters an obstacle that is either to the right or left of the line. The child must first call out to which side of the line the obstacle is before either crawling under it, over it, or going around it, etc.

BALANCE

1. *Object balance:* Carrying a beanbag or basket on their heads, the students are directed to walk forward, backward, sideways, or twisting or turning. *Variation:* Changes of speed using various tempos of music. Direct the students to find another part of the body on which to carry the beanbag.

2. *Object balance:* Walk, run, or walk blindfolded while carrying a marble or small ball on a spoon. *Variation:* Interlace fingers together, holding palms down and away from the body, and perform various locomotor activities while holding marble balanced on top of hands. Two children face each other and form a "tray" by putting their four hands together, backs upward. They try to carry a marble on the "tray" as one walks forward, the other backward.

3. *Dynamic balance, strength in leg muscles:* Stretch arms overhead and clasp hands together and walk on tiptoe with legs stretched and knees straight.

4. *Dynamic balance, coordination, agility:* Walk, run, skip, jump or gallop to the music. When the teacher stops the music, students may pose any way they wish, but only one foot may touch the ground. After a count of five, the teacher starts the music again and the game continues.

5. *Perception of shapes, letters or numbers, dynamic balance, agility, strength in leg muscles:* Place large squares, discs, or triangles on the floor close together. The students step, jump or hop from one "stepping-stone" to another. Gradually increase the distance between them. The children should call out the name of each shape as he steps on it. *Variation:* "Stepping-stones" may have numbers or letters marked on them. The children should call out the letter or number as he steps on it.

6. *Dynamic and static balance:* The students may turn round in whirling motions sufficiently controlled to enable them to achieve a static position when the music stops. At this signal, they jump in the air, land in any position they choose, and remain still. Direct the students to experiment with this activity in teams of two or in groups of three or four.

7. *Balance tag:* Book (or beanbag) on head. At signal start walking, try to tag one ahead—anyone tagged or losing bag off of head is out of the game.

FLEXIBILITY

1. *Spine and leg muscles:* Swing arms and upper body downward in any fashion and swing up, using the momentum of the

upward swing to jump or leap forward. Progress around the room to music or use as a relay race.

2. *Knee joints:* Hold calves or ankles and walk forward, backward or sideways. Caution: students with hyperextended knees should keep knees relaxed or slightly bent.

3. *Spine, coordination of arm and leg movements, dynamic balance:* Elephant walk, link the fingers of both hands, bend forward at the waist and letting arms swing loosely, walk forward with heavy steps.

4. *Spine, strength in abdominal muscles:* Forward and backward somersaults.

5. *Hip joints, strength in trunk muscles, static and dynamic balance:* Begin in kneeling position sitting on heels, back straight, fingers laced behind neck. Bend to left and right without leaning forward. Progress to upright kneeling position and repeat. Progress to kneeling on one knee with opposite leg stretched straight out to side and repeat bending movement. *Variation:* In standing position with one toe pointed to side. Progress to walking on toes while turning, twisting and bending.

COORDINATION

1. Walk on hands and feet with legs and arms straight or bent, arm and leg on same side moving together. Repeat with one foot off the floor, or one hand. Repeat using opposite arm and leg. *Variation:* Use blindfolds and direct students to move toward the sound of your voice or other sound cues as you move around the room.

2. *Dynamic balance, strength in shoulders, trunk and buttocks:* Students assume position for crab-walk and when in raised position, turn over so that they face the floor. They must not touch the floor with any part of the body except the hands and feet.

3. *Strength in leg muscles, speed and agility:* Students practice jumping forward holding beanbag or ball between their knees or between their feet. Can be used as a relay race or as a game with one less beanbag than the number of children. Place beanbags in

Locomotor Exercises

the center of the room and at a given signal the children race to get the beanbags. The one who fails to get the beanbag must collect all the beanbags and return them to the center so that the game can be played again.

4. Walk, run or change directions while throwing beanbags or balls in the air and catching them.

5. *Walk the dots:* The teacher puts sixteen discs on the floor in four rows of four discs each. Each is given a card on which a plan of the discs is drawn and a route is marked. Each walks or jumps from disc to disc following the route marked on his card.

EXAMPLE:

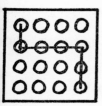 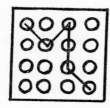 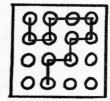

AGILITY

1. Kneel on the floor and rise to a standing position as quickly as possible, using one hand if necessary. Repeat but with arms folded so as not to use hands for assistance. Repeat beginning in kneeling position with toes bent forward and rise to standing position quickly without using hands.

2. *Strength in arm muscles:* Lie face down, hands beside shoulders, palms on floor, elbows bent. Jump to a standing position as quickly as possible.

3. *Body awareness:* Sit on floor, knees bent, feet flat. Rise to standing position as quickly as possible, using one hand if necessary. Progress to rising from a sitting position tailor-fashion, legs crossed. Two children may do this as a team, holding hands, standing and sitting together, giving each other support. Progress to having students sit back to back, arms linked, feet flat on the

floor. Rise to standing position by pushing against each other and return to original position.

4. Lie on back with arms stretched overhead, roll over sideways in any direction, curl up with knees and elbows bent, head tucked down as close to knees as possible and roll in this position. Return to stretched out position continuing to roll and back to curled up position and repeat.

EXAMPLE ONE: For teaching running, turning, leaping, agility and space.

"Race You Down the Mountain"

Record: Woody Guthrie (*Children's Songs*) Golden Records LP #238

"I'll race you down the mountain, I'll race you down the mountain, I'll race you down the mountain, we'll see who gets there first."

Movement: Running around the room in large circle, single file.

I'll chase you 'round the bushes," etc.

Movement: Running in circle using personal space.

"Let's run and jump the river," etc.

Movement: Run and leap around the room in large circle

"I hear myself a huffin', a huffin', and a puffin'," etc.

Movement: Deep breathing while running slowly and heavily

"Let's rest beside the water," etc.

Movement: Sit and rest while continuing deep breathing.

"Now! I'll race you down the mountain," etc.

Movement: Everyone must be standing on the word "Now! and repeat running as fast as possible, taking care not to bump into anyone.

EXAMPLE TWO: For teaching crawling, rolling, crabwalk, and concepts of heavy movements.

"L' Elephant"

Record: Leonard Bernstein (*The Carnival of the Animals*) Columbia Records ML #5768

Movement: Crawling from one side of the room to the other using heavy movements. Gradually add rolling and the crabwalk.

Locomotor Exercises

EXAMPLE THREE: for teaching a variety of locomotor movements and creativity.

"Dance Around"

Record: Woody Guthrie *(Children's Songs)* Golden Records LP #238

"Dance around and around and around and around," etc.
  Movement: Students may experiment with various movements in a circle.

"Hold up your hands, your hands, your hands," etc.
  Movement: Still moving in circle students can experiment with movements while holding hands above head.

"Walk on your toes, your toes, your toes," etc.
  Movement: Still experimenting while walking on toes.

"Now march and march and march and march," etc.
  Movement: Same as above while marching, lifting knees high.

"Now take big steps, big steps, big steps," etc.
  Movement: Same as above.

"Hold up your hands and take big steps," etc.
  Movement: Follow the words and music.

"Dance around and around and around and around," etc.
  Movement: Repeat as in the beginning.

EXAMPLE FOUR: For teaching both creativity and imitation.

"Going to the Zoo"

Record: Peter, Paul and Mary *(Peter, Paul and Mommy)* Warner Brothers Records LP #1785

"Daddy's takin' us to the zoo tomorrow," etc.
  Movement: Walking or skipping in circle while clapping.

"See the elephants," etc.
  Movement: Imitate elephants.

Chorus: same

"See all the monkeys," etc.
  Movement: Imitate monkeys.

Chorus: same

"There's a big black bear," etc.
  Movement: Imitate bears.

Chorus: same

"Well the seals in the pool all honk, honk, honkin'."
   Movement: Imitate seals.
Chorus: same
"Well we stayed all day and I'm gettin' sleepy."
   Movement: Students yawn and stretch arms above head gradually relaxing from the head down to the toes. Get down on one knee, then on the other knee, sit to the side, yawn and stretch once more and lie down relaxed and still.

NOTE: The song continues into the chorus again but it is suggested that the music be faded out by adjusting the volume, thus maintaining control of the class for the next exercise or academic lesson.

Chapter 20

Dance Composition

When the teacher is composing a dance or series of movements as in the examples in chapters 18 and 19 the following steps should be followed:

1. Choose proper music to fit the age group and the abilities of the students. You will find many usable folk songs and musical rhymes suitable for elementary grades. Contemporary music is the key to "turning on" junior high school students and high school students. All must be exposed to the classics but with a light touch. Listen to the music several times and make a note of the number of counts in each measure, the measures in each phrase, phrases in each section, the timing, rhythm and tempo of each section. Decide on a mood or theme that the music suggests.

2. Choose an overall movement pattern and floor pattern. The music may suggest skipping, jumping, turning, clapping or flowing, balance, posing or swing movements. Decide if the dance should be done in a circle, line, small scattered groups or with partners. If the students begin in a circle, can you direct them into a line, diagonal line and back into a circle? Make a note of all your ideas. You can tighten and make changes later, but one idea will lead to another. Some ideas which seem good may turn out to be wrong for this particular music or that particular group of students, but you can file them away and use them another time.

3. Group students according to size and ability. All your students will be able to perform well if you have made certain to use only movements and steps which they have learned successfully. Grouping can be used to create interest if the entire class performs one part, while, as the dance progresses, separate groups are used to perform different movements alternatingly, e.g., as

parts of an orchestra play various parts of a musical composition.

4. Plan the choreography to fit the music, movement pattern, floor pattern and students. This must be done before class using steps that are familiar. You will probably add some steps and discard others. Advanced students are challenged by changes, whereas beginner students might be confused by changes. It might be safer and give your students more security if you stick to your original plan, even if you decide it's not good. When your students are invited to perform in public you will have a repertoire of your best compositions ready and the bad ones can be discarded.

5. Plan entrances and exits according to your music and space. Will your students wait for a short introduction of music and then dance to their places on the stage or will they enter and pose until the music begins? Will the music fade out as they exit or will they execute a pose at the end of the music and then exit? Your students will be more secure and appear more polished if you plan each detail well and give definite directions.

You can get many of your ideas from watching dancers on television, attending live dance concerts or local dance recitals. One teacher I know went to see "West Side Story" several times, with a flashlight and a notebook writing down the dance movements step by step. Props such as tambourines, maracas, balls, ribbon streamers, or jump ropes can inspire creative ideas. There are many books available that can be invaluable if you are not the creative type. Records are also available featuring spoken classes to be used in conjunction with matching exercise manuals. These are excellent as far as they go, but I feel they should only be used to get you started or to ease you through a period when your own ideas temporarily run dry. You can be sure that every one of your students will each day have a chance to create original activities.

Do not make the mistake of telling your students exactly how to move and then labeling this creative dance movement. If you have created something for them call it a routine. If you wish them to create, first give them an adequate movement vocabulary and then let them do the creating, the thinking.

Use the following suggestions:

1. *Personal space.* Students are diected to stand as far away

Dance Composition

from each other as possible so that they will not hinder each other in moving. They try to find out how far they can extend their arms and legs in all directions. Compare the child's personal space to a bubble surrounding him. Another way is to have the child pretend he is inside a giant glass jar or pop bottle and his creative ideas for various movements are pouring in through the top of the bottle.

2. *Common space.* The students walk, run or skip, etc., to music around the room trying to find out how far they can go in all directions, avoiding bumping into each other, the walls of the room or the furniture. The teacher explains the room has to be shared by everyone; it is called common space.

3. *Floor patterns in the common space.* The students are directed to walk, run, skip, etc., describing different floor patterns while always facing the teacher, a wall or a particular object (focus). A rope may be used to form a floor pattern. The children may move in straight lines, curves, sideways, backward, zigzags, circles, etc., all without changing body direction.

4. *Floor patterns while changing body direction.* Same as above except that each child may now change body direction while describing floor patterns, going backward, forward, sideways, etc.

5. *Shapes.* Students explore various shapes they can make with their bodies. They find out how their bodies can be narrow, wide, flat, round, twisted, etc.

6. *Air patterns.* Students move head, arms, legs in different directions, bend their trunks and move their bodies describing patterns in the air in as many ways as they can devise, using only their personal space. *Variations:* Students can use balls or ribbon streamers to describe patterns in the air.

7. *Changes in tempo.* Move in a circle, running, walking, skipping, etc., changing from slow to quick movements. Alternate tempo frequently.

8. *Changes in weight.* Practice walking or running heavy steps, then light steps. Make a big jump. Alternate one heavy step with three light steps. Direct students to create their own combinations of heavy and light steps.

9. *Changes in body support.* Students experiment with supporting their bodies on one, two or more points, walk on three or four points, roll on the ground, hop on one foot, etc. Practice transferring weight from one part of the body to another, e.g., changing from standing to kneeling to standing to hands and knees, etc.

10. *Level.* Change from high level on tiptoe, to medium to low level with knees bent while moving around the room.

11. *Flow.* Alternate continuous flowing movements with those requiring immediate stops in order to demonstrate what flowing movements mean.

The more advanced the student, the more he enjoys improvising. Improvisation contributes to the development of the individual personality. It helps to free him from the physical and psychological inhibitions which plague so many children and adults. Improvisation acts as an outlet for release of emotions, discovery of new ways of moving, new combinations of parts of the body, discovery of self and a new vehicle of self-expression.

Improvisation takes time. Ordinarily there is not enough time during a class period to work out a movement or dance composition until it is polished.

Onlookers will think the whole thing quite chaotic. Only those involved can feel the benefits accrued by improvisation. Students will not verbalize what they feel, but they *will* feel! Pressure will be brought on a teacher to produce a composition that looks orderly, with students all in a row, moving in unison like little tin puppet soldiers. This may be quite pleasing to the spectator but does not do much for the development of an individual. It is the teacher's responsibility to maintain a proper balance between techniques taught as exercise drill and improvisations used as a tool of personality development. Do not throw discipline out the door just to let creativity in the window.

Creative dance can be neither good nor bad. If a student uses a movement that describes how he feels, who can say he is wrong? If the movement is pleasing to the spectator it may be due to the fact that the student has overcome the obstacle of body control

Dance Composition

and balance and is able to project his feelings to others. He has developed kinesthetic awareness, the relationship of the various parts of the body, and the need for skill in control and coordination. Finally students should experience dance as dance, not as an exercise or as preparations for dance.

HINTS FOR IMPROVISATIONS

1. Walks, skips, or slides in a count series of eights, fours or twos. It is suggested that the beginner student use only counts or series of eights until he attains an amount of mental and physical agility. Slides might begin in a circle with students first facing into the center and then out. Students may choose to move with partners, proceeding to the right and then changing directions. The use of high and low levels can make monotonous steps more interesting.

2. Floor patterns promote variety. Add interest by using a zigzag pattern, a diagonal, a square or use two opposing groups moving away from each other and coming together again.

3. One or two focus points can be used to give more of a polished performance. Students might perform any number of walks, turns, jumps or bending movements while focusing eyes on one point or object in the room.

4. Sounds of waves, birds, thunder, wind, animals, motors, bells, trains, whistles, horns or sirens can suggest movement. Interesting experiments can develop with recorded sounds when students are not told in advance what the sound is.

5. Children's rhymes and folk songs for elementary students can present ideas for new movements. High school students and adults having a larger movement vocabulary will respond to more sophisticated poems and prose.

6. Words such as *silk, slime, velvet, sawdust, sand, mud* or *fur* will get ideas started. Emotional words like *suspense, sorrow, grief, joy, ecstasy, indecision, frustration, worship, anger, terror* or *agony* will tend to bring out acting abilities of your students as well as serving to release anxieties, especially when accompanied

by appropriate music. Names of animals are excellent for inspiring elementary students. The freer the student, the easier it will be to express emotion through movement.

7. Other ideas to help creative thinking may be: blindfolding students and directing them to move about the room using only tactile and auditory stimuli; use patches of light or shadows on the classroom floor to move around or between; use pictures, fabrics, colors or shapes to stimulate ideas.

EVALUATIONS

This is a vital part of the creative dance experience. The student wants to know how to improve his work but he also needs to feel successful. The student must feel free to experiment and to make mistakes. For important works or for grading puposes, the teacher may wish to use the following items as a guideline. The students may also make use of the list for self-evaluation. As with evaluating any creative work take into consideration that each person involved will have a right to his or her own individual opinion.

1. Did the composition use a large or small space according to directions given by the teacher or suggested by the music?
2. Did the composition use at least three changes of direction?
3. Were at least two variations in levels used?
4. Were both stationary (sustained) and locomotor movements used?
5. Were at least two focus points used?
6. Did the composition have an idea or theme?
7. Was the idea well expressed?
8. Did the dance have a beginning, middle and end?
9. Was the composition of an appropriate length?
10. Did the students use any new movements?
11. Did the student use the whole body as a tool of expression?
12. Were the transitions from one part of the dance to another done smoothly?

Dance Composition

Whatever form of evaluation is used it is important to keep a daily, weekly or monthly record of each student's progress and to remember that each student is competing only with himself. The final evaluation should show that the student has grown in technique, creativity and self-concept.

Chapter 21

Therapeutic Exercises

EXPLANATION OF TERMS

1. Abnormalities of the spine
   a. *Lordosis* (saddle back or swayback): abnormal curvature in the small of the back accompanied by weak abdominal muscles and in some cases hyperextended knees.
   b. *Skoliosis:* a lateral curvature either C-shaped or S-shaped causing one shoulder to be higher than the other, or one hip higher than the other.
   c. *Kyphosis angularis* (humpback): a severe curvature of the upper back requiring bed rest and special medical care.
   d. *Kyphosis acquired* (round shoulders): a slight curvature of the upper back caused by weak muscle tone in the upper back, shoulders and chest.
   e. *Kypholordosis:* a combination of swayback and round shoulders most complicated to correct since the improvement of one may exaggerate the other.
2. Abnormalities of the leg
   a. *Genu recurvatum* (hyperextended or swayback knees): an abnormal curvature of the leg caused by weak or overstretched ligaments behind the knee joint.
   b. *Genu varum* (bow legs): easily detected by directing the student to stand with feet together and toes pointed straight ahead. The feet touch but the knees do not.
   c. *Genu valgrum* (knock-knees): detected by asking the student to stand with feet together if possible and toes pointed straight ahead. The knees touch but the feet do not. Many obese students will have the same difficulty but this is due to overweight and is not evidence of bone

Therapeutic Exercises

deformation. Knock-knees are usually accompanied by flat feet.
   d. *Tibial torsion* (cross-eyed knees): detected by asking the student to stand with feet together and toes pointed straight ahead. You will notice that one or both knee caps do not face straight ahead but face inward.
3. Abnormalities of the feet
   a. *Pigeon toes* (turned in): may be combined with rolling out of the feet and caused by weakness and overstretching of the muscles on the outside of the foot.
   b. *Turned out:* usually combined with rolling in and flat feet. Caused by weakness and overstretching of the muscles on the inside of the foot.
   c. *Flat feet:* the absence of a proper arch due to weakness of the muscles and resulting eventually in bone deformation. All toddlers normally have flat feet.
   d. *Rolling* (inward or outward): caused by failure to use the foot properly and to maintain three points of contact with the floor, big toes, little toe and heel.

When a student is placed in your class, it is your responsibility to train both his mind and his body. You must first check the physical build of the child and look for any physical defects. If there is any question as to the physical condition, consult with the school administrator, parents, or family physician. The lesson must be planned with the purpose of eliminating overstrain and emphasizing a balance and variety of activities. Each movement must include tension and relaxation of the involved muscles. If body placement is correct there will be less strain, thus insuring less fatigue.

No amount of training will correct a deformity of bone structure. With adequate movement education the general body appearance will be improved and poor bone structure can be camouflaged. Abnormal bone structure may have been caused by disease such as rickets or Pott's disease, tuberculosis or polio, incorrect posture, malnutrition, congenital dislocation of the hip, weakness of abdominal or back muscles, rapid growth, walking at too early an age, or overweight. Improved body placement, carriage, breathing habits, balance and coordination can do much to disguise defects.

Kyphosis acquired (round shoulders) is caused by abducted (forward tilted) scapulae. Often teenagers, especially those who have experienced rapid growth, show a degree of this condition because of poor posture habits and failure of the scapular adductor muscles to hold the shoulders in place.

The "relaxed" type of round shoulder, can be corrected with exercises to stretch and strengthen the proper muscles. The "rigid" or acute round shoulder is more difficult to correct but can be improved through the use of exercises to stretch and lengthen the pectoral muscles and strengthen the trapezius muscle. Exercises in a face down position enable the student to stabilize the thorax while isolating the movement of pulling the shoulder blades together (adduction of the scapula). Whenever you become aware of tension in a student, stop and show him how to relax physically.

Teachers should be aware that any incomplete movement can create bulky muscles. Any movement which requires a contraction of the muscle must be followed by relaxation. Contractions, rapid movements and tension tend to pump or blow up muscles till they become inflexible.

Authorities agree that failure to relax muscles is a factor that makes muscles become hard and inelastic. The alternating of muscle stretching and relaxing is also a safeguard against strain, and cramps especially in the foot muscles and in the thighs.

Exercises should never be executed in a jerky manner. They should never be rushed. The student should never be embarrassed by a teacher's remarks or by comments that point to his condition. Too often, complexes, anguish, and disillusion have already left their psychological mark. The most sensitive part of any human being is his ego. "The good teacher is the one who knows how to encourage progress—not discourage it." Pointing out improvement will reaffirm a student's faith and confidence in his teacher as well as encourage the effort and discipline necessary for this progress.

Teachers must know that some physicians will give a student

a "blanket excuse" from all physical activity. Not necessarily because any activity is harmful, but because the physician may be unaware of the available adapted program. Many times these students are the very ones most in need of developmental activity. Always question a "blanket excuse" and inquire as to what is or is not permissible during physical activity class.

EXERCISES

(designed to improve body mechanics and prevent injury)

1. "Out of Sight." Lie on back with feet together and toes pointed. Press shoulders down against the floor and also lower curvature of the back, feeling the spine resting as flat on the floor as the rest of the body. In eight counts, keeping the legs straight and the toes pointed, lift the feet off the ground until they acquire a 90 degree angle with the body. At this point, starting the movement from the tailbone, try to feel an inch by inch rolling of the spine against the floor as the toes reach to touch the floor over the head. Remain a few seconds in this position to reverse the circulation of the blood stream and relax pressure on the legs before starting the return. Beginning from the neck, roll downward in the same slow, even rhythm, press the spine down little by little, until the tailbone and then the legs have reached the initial position.

Many students may find this exercise difficult to execute at first—especially those who have a stiff spine and shortened dorsal muscles. These students should not try to reach all the way back immediately but, in practice, try to reach a little more each time. The spine will eventually stretch until the maximum reach can be accomplished.

A particular effort to relax should be made while doing the exercise. Its rhythm should be smooth, without jerky or rough pushing. It's advisable that a teacher assist at the beginning until the student has confidence and has built up weak muscles. Care should be given not to apply force while assisting, for the forced action can develop fear in the student—and cause nervousness or panic, the result of which is that the muscles contract.

Sometimes a very bony type of spine bruises during this exercise. A mat or piece of carpeting will prevent this. Never use a cushion or anything with a springy surface.

Knees should be straight during this exercise for maximum benefit. Arms should always remain relaxed against sides of the body.

By doing this exercise two or three times a day, a student with a weak spine can improve its pliability and suppleness. The exercise stretches all the muscles of the back and the intervertebral ligaments. It eventually helps, to some degree, swayback and other imperfect curvatures of the spinal column.

2. "The Rolling Stone." For the student who has difficulty in the "Out of Sight," there is a simple rolling exercise which helps to strengthen and stretch the spine, helps to increase movement in a spine stiff from nervous tension, calcium deposits, or arthritis.

Sit on the floor with the knees bent, arms clasped around legs, head forward touching the knees. Maintaining this position, roll backward and forward several times, being careful not to rest your weight any place until the end of the exercise. Working specifically with the pelvic girdle where the body is most often disabled through injury or ailment, the following exercises are created to increase movement in the hip joint and to stretch the thighs. The main concern is with increasing range of motion and flexibility through both outward and inward rotation.

3. "Going in Circles." Along with increasing strength and flexibility, also strengthens the muscles of the abdomen. Lie on back with legs apart and extended, toes pointed. Lift one leg several inches from the floor and rotate leg describing a small circle in the air, repeat and rest. Repeat exercise with the opposite leg, following with circle movement in reverse. Make certain that the movements are isolated in the hip joint, keeping the entire body flat, especially the lower spine. Progress to lifting both legs and executing movement with both legs at the same time.

4. "Half Way." Lie on back, spine flat, shoulders down, and legs as far apart as comfortable, arms resting at sides. Without moving the thigh and keeping the knees as close to the floor as

Therapeutic Exercises

possible, slowly move the feet toward each other until the toes meet, thus creating a "diamond shape." Slowly return to starting position. Progress to lifting and lowering the pelvis while legs are in diamond-shaped position, still keeping knees close to the floor.

5. "Puff and Stuff." Lie on back, legs extended and together, shoulders and lower back pressed against the floor. Flex the right knee, grasp knee with both hands and bring it toward the chest. Keeping hips and lower back pressed down, pull knee across body to left, then to the right as far as possible. Progress to describing a circle four times in one direction and four times in the reverse direction, keeping knee bent throughout. Repeat with left leg. To avoid strain, be careful to bring the knee toward the midline of the body in each circle. (Physical therapy departments of many hospitals apply this exercise in a passive form to rehabilitate patients with chronic conditions of the hip joint.)

6. "The Bugaloo." *Caution:* this exercise must not be done until students are sufficiently warmed up. Lie on back, knees flexed and feet apart and as close to the body as possible, flat on the floor. Bend right leg toward left as far as possible, and bend left leg toward right, squeeze muscles of thighs as if you were holding a piece of paper tightly between them. Return to starting position and change legs and repeat. Counting: count 1—bend right leg, count 2—bend left, count 3—squeeze, count 4—return to starting position.

A student with bow legs usually has weak hamstring muscles (back of the thigh) difficulty in locking the knee joint and in using adductors (muscles in charge of tightening and holding the legs together toward midline, and often has an exaggerated sagging of the gluteal muscles.

7. *Exercises for bow legs:* Stand, holding a chair or table for balance with one hand, or with palm flat against the wall (for better control), heels together, toes out slightly. Keeping the knees straight and placing the body correctly, tilt the pelvis back and then slowly pull it forward tightening all the muscles of the inside of the thighs, bringing the knees as close as possible. Release the pelvis, returning to original position and repeat exer-

cise a number of times. If faithfully pursued, it will achieve a general improvement of appearance in which the irregularity definitely becomes less conspicuous.

8. *Exercise for "cross-eyed knees"* (tibial torsion): Lie face down, head resting on folded arms, legs straight and together, toes together and heels apart. Bring heels together with a slow, strong contraction of the buttocks. Hold for a few seconds, then relax slowly. Repeat action six times.

9. *Exercise for relieving leg cramps and stretching the Achilles tendon and the calf muscle:* Sit on floor, legs together and straight, and toes pointed. Reach forward with hands and hold outside of feet. Pull toes toward body, relax and repeat.

10. *Exercise for stretching calf muscles, etc., as above:* Facing the wall but standing several feet away from it, lean forward placing hands on wall, arms straight, feet together, knees straight. Bring one foot forward, knee bent, stretch the back muscle and tendon of the opposite leg pressing heel down with small bouncing movements. Alternate legs.

11. *Same as above:* Standing with balls of feet on edge of wood block or book, lower heels to the floor and slowly rise halfway to tiptoe position. Repeat smoothly several times. If one calf muscle is shorter or less developed than the other, exercise only the underdeveloped leg.

12. *Exercise for alignment and control of the lumbar spine and pelvis (body control and placement):* Stand slightly away from the wall. Feet two inches away. Lean back, flex knees, press both small of the back and neck flat against the wall. Slide trunk up the wall keeping spine flattened against it. Return to standing position with knees straight (not hyperextended) maintaining identical spine position away from wall. Lean body forward so that weight is over balls of feet. Most students push away from wall using buttocks, thereby loosing placement or attempt to keep back and legs stiff. This exercise can be adapted so that beginner students can do this exercise on the floor until they get the feel of pressing spine against the floor.

13. *Stretch for hamstring muscles:* Lie on back, legs together and straight, toes pointed. Flex right foot, stretch or slide right

leg and hip downward and interlace fingers behind the knee. Without jerking, press the leg gently toward the chest (in a springlike action) until the "pain" of the stretch can be felt along the back of the leg. Release. Repeat springlike movement several times before returning to original position. Alternate legs. Repeat four times with each leg. This exercise is especially safe for students with hyperextended knees.

14. *Exercise to stretch hamstring, quadricep and gluteal muscles and reduce gluteal muscles:* Lie face down, legs together and straight, toes pointed. Bend right knee and reach back with left hand, clasp foot and stretch thigh off floor as far as possible. Repeat with opposite foot and hand. Progress to clasping both feet (same hand to same foot) legs slightly apart and rock on abdomen and pelvis. Check that both thighs are off the floor.

15. *Flexion and extension of knee joint:* Lie on back, legs together and straight, toes pointed, shoulders and spine touching floor. Flex knee and bring to midline toward chest. Press both hands over knee. Extend leg, lace fingers behind knee and pull toward chest. Flex knee and repeat.

16. *Stretch for hamstring muscles and sciatic nerve:* Stand with arms stretched overhead, right foot forward, left back, knees locked. Bend forward and reach arms around both legs, clasping hands behind left knee. Bounce several times and return to starting position. Alternate legs.

17. *Stretch for quadriceps muscle (and to relieve "Rider's Cramp" in the groin area):* Stand in second position, feet apart, knees locked, hands on hips. Flex left knee, do not tilt hip or raise heels from the floor. Knee must be flexed directly over toe. Progress to increasing stretch by lifting heel of flexed leg, shifting weight onto the ball of the foot. Repeat to other side.

18. *For developing the arch of the foot:* Stand with legs apart, feet parallel. Roll weight to outer border of foot and then to inner border. Relax and repeat several times.

19. *For flexion and extension of the ankle:* Lie on back, legs together and raised, toes pointed. Flex and extend feet alternately.

20. *For strength and flexibility of the ankle and to stretch Achilles tendon:* Lie on back, legs together and straight. Lift right

foot off the floor and make a complete circle as you rotate foot from ankle outward four times, inward four times. Repeat with other foot.

21. *For stretching muscles of the foot and leg:* Stand with feet together, right foot on ball of foot, knee flexed, left heel down, leg straight. Using dynamic tension, slowly press right heel down and at the same time flex left knee and raise left heel, relax and repeat.

22. *For strengthening short muscles of the foot and to improve arch:* Stand with feet apart, draw toes up so that only the heels and tips of toes touch the floor. Release toes, relax foot and repeat.

23. *Stretch for thorax and neck muscles:* Sit with knees flexed, feet flat on floor, arms straight at sides with palms pressing floor, back rounded, forehead touching knees. Working in isolated sections, start by stretching chin forward and upward, next stretching neck, then chest and abdomen, until reaching maximum stretch. Return to original position and repeat.

24. *To stretch pectoral muscles and strengthen muscles of upper back:* Lie face down, shoulders relaxed and touching floor, clasp hands in back. Lift head and chest, pulling arms backward until shoulder blades are pressed together. Relax and repeat.

25. *For mobility of shoulders and to relieve tension in neck and shoulder area:* Lie on back, shoulders relaxed, arms at sides, palms down, knees flexed and feet flat on floor. Rotate shoulders forward, upward and downward in smooth continuous movement. On the downward movement, press shoulders to the floor as much as possible. Relax and repeat several times.

26. *Resistive exercise to strengthen muscles of the upper back and shoulders:* Sit with legs crossed, back straight. Clasp hands behind head and using force with right hand pull left hand in opposite direction, resisting as in dynamic tension. Repeat several times.

27. *For mobility and relief of tension in neck and shoulders:* Rotate head.

28. *To stretch muscles of shoulders, chest and back:* Airplane exercise. (On stomach, raise arms and legs off floor, stretch, relax and repeat.)

Therapeutic Exercises

29. *To strengthen abdominal muscles and stretch lower back:* Lie on back, arms up with elbows bent and backs of hands and elbows touching floor, legs together. Flex knees toward chest as far as possible, keeping toes pointed. Contract abdomen and press lower back against the floor. Lower legs slowly (still bent) until toes touch the floor, maintaining contraction. Extend legs slowly until completely straight.

30. *For strengthening abdomen and reducing waist:* Rotate body.

31. *To stretch and strengthen abdominal muscles:* On hands and knees with head up, contract abdomen, arching back and bring head down. Return to original position and repeat several times.

32. *To strengthen abdominal muscles:* "Clown Walk." Stand with hands clasped in back, arms stretched as far as possible. Lift one knee and try to touch forehead, bending supporting leg slightly. Alternate legs and repeat several times.

33. *To strengthen upper abdominal muscles, upper back and neck:* Lie on back, arms up with elbows bent, backs of hands and elbows touching floor, legs straight and together. Stretch chin forward, stretching neck and bring chin as far as possible toward chest. Relax and repeat several times. Teachers working with young children say, "Look at your toes."

34. *To improve breathing and body awareness:* Stand with head drooping. Breathe in steadily while raising the arms and head. Breathe out steadily while returning to original position and repeat. *Variations:* repeat while sitting tailor fashion or kneeling.

35. *Same as above:* Lie on back with hands on abdomen. Breathe in and out while feeling the abdomen rise and sink in continuous rhythm.

36. *Same as above:* Lie on back and steadily breathe in as you slide arms upward along the floor until they meet overhead. Breathe out steadily as the arms slide to original position. Repeat.

37. *For strength in leg muscles and flexibility in ankles:* Sit on floor and practice foot movement as used in jumping, e.g., bend feet and toes toward body then point toes away from body as far as possible. Practice making small jumps pointing toes

toward the floor. Take precautions to make sure the knees are bent prior to leaving the floor and after landing and that heels are pressed down at the completion of each jump.

38. *Thigh muscles:* Kneel in upright position, slowly sit back on heels then rise to upright position and repeat. May be done in pairs with children facing each other and holding hands to give support.

39. *Leg and abdominal muscles:* Lie on back, legs straight, bring one knee as close to chest as possible, alternate legs. Repeat while holding straight leg slightly above the floor.

40. *Thigh and trunk:* Lie on side and raise leg sideways either straight or bend knee and then straighten. Alternate legs. May also move legs in circular motion.

41. *Lower back:* Lie face down, arms on floor above head. Slowly lift arms, head and chest as high as possible then slowly lower. (Forward bending should follow exercises in which spine is bent backward.) Persons with swayback should avoid this exercise. Repeat raising legs and keeping upper body on floor.

42. "Seal Walk." Shoulders, arm and back muscles, coordination and object balance: Lie face down weight on hands, arms straight, shoulders and upper body lifted. Walk forward on hands dragging legs and feet behind. May be done with beanbag or basket balanced on head.

43. *Arm, back and abdominal muscles and flexibility:* Bend forward, placing hands on floor as near to feet as possible, legs straight. Walk forward with hands until body is outstretched, then either walk hands back to original position or walk feet forward to meet hands. For a good stretch in calf muscles and Achilles tendon, heels must be pressed down throughout exercise.

NOTE: Watch for abnormal breathing. Without proper breathing, no exercise can be of any help. Proper breathing is a source of vitality and its neglect often leads to shortness of breath, sunken chest and poor circulation. Classrooms must be clean and well ventilated and teachers should be especially careful that no drafts are directly over the students. Proper ventilation is determined by the number of students in the room, the size of the room, the

Therapeutic Exercises

number of lights, radiators and types of activity involved. In winter students should be well covered until the body is warmed up and can provide its own heat. Before and in between exercises, students should be directed to breathe in through the nose, to fill the lungs with oxygen, holding for a few seconds and then exhaling slowly through the mouth. This simple, preparatory breathing exercise will educate the student to breathe normally. Opening their chests and expanding their lungs to keep more oxygen in reserve, they'll later find jumps and complicated routines easier to do.

Physicians have found that some thyroid problems and enlarged adenoids can be aggravated by breathing through the mouth. More severe illnesses such as tuberculosis, asthma, stomach and circulatory problems can also reflect this poor breathing habit.

Upper, middle and lower breathing: upper breathing is shallow and only the highest and smallest part of the lung is used, allowing very little air into the lungs. Middle breathing is still not the most efficient. Lower or abdominal breathing is the most complete and the entire lung is filled with air.

NOTES ON POSTURE: The student with the best posture will progress faster, reach a higher degree of attainment, and last longer. Many "experts" believe that posture can be improved by exercise and by holding parts of the body in correct positions. Such practice may result in better looks temporarily, but both flexibility and efficiency have been sacrificed.

Flexibility is limited in the poorly aligned body. More muscle power must be applied to overcome the restriction in movement by tightened muscles. This may lead to muscular overdevelopment with distorted body contours, especially in the thighs and calves.

Body mechanics, therefore, must emphasize relaxation and the value of freedom of movement. With prolonged poor training during early life, poor body alignment and muscular problems become so deeply ingrained that no amount of good teaching can repair the damage done, especially at the knees, ankles and feet.

Chapter 22

Jazz Eurythmics

The following exercises have been found to be most beneficial for adolescents, teens and young adults. The movements serve to develop coordination and strength, using the whole body to its fullest capabilities. The movements are compatible with the music of today.

WARM-UP

1. Stand with feet apart, arms down at sides, pull in stomach and tighten hips as you reach up with the right hand, stretching as you reach (2 counts). Repeat with left hand, right and left again (2 counts for each stretch). Bend forward and down to touch toes (4 counts), and up to maximum stretch, ready to repeat exercise (4 counts). Repeat complete exercise four times in all, then repeat touching toes only, four times. *Variation:* bring feet together for last four toe touches.

2. Stand with feet apart, arms down at sides. Raise shoulders (2 counts), lower shoulders (2 counts), repeat for a total of 32 counts.

3. Stand with feet apart, left hand on hip. Stretch right arm up (2 counts), bend right elbow bringing hand toward shoulder (1 count), and push right hand out to side away from shoulder (1 count). Continue until movement is completed four times in all and then repeat exercise with left arm.

4. Stand with feet apart, raise both arms overhead and push hands, palms down toward floor (4 counts). Repeat four times in all.

5. Stand with feet apart, hands on hips and bend from the waist to the right and to the left (4 counts). Continue for 32 counts.

6. Stand with feet together, arms overhead, shoulders back and down, pull in stomach and tighten back and hips. Bend knees until you are almost sitting on heels, bringing arms together in front and down (4 counts). Rise to standing position bringing arms up the sides to original position (4 counts). Keep back straight throughout. Repeat four times (32 counts).

7. Stand with feet apart, arms out to sides shoulder high, tighten hips, pull in stomach and expand chest. Bring the right shoulder forward and the left back (2 counts). Reverse and repeat for 32 counts. Arms do not move.

8. Stand with right foot in front, left back, feet apart. Bend elbows bringing hands in front of chest. Raise hands to overhead position (4 counts). Turn palms out and push hands down and out to sides shoulder high. Tension should be felt from fingertips to shoulder blades (4 counts). Repeat four times (32 counts). Repeat four more times raising arms on the count of one, hold two and lower arms slowly on counts 3 through 8.

9. Stand with feet apart, arms overhead. Keeping knees straight drop body forward and swing arms through legs and bounce twice. Return to original position (8 counts). Repeat four times (32 counts).

10. Stand with feet apart, arms out to sides, shoulder high. Slowly bring right arm overhead, across body to left and back to original position (8 counts). Repeat with left arm, right and left again (32 counts in all). Repeat complete exercise again with 4 counts to each movement (16 counts).

11. Stand with feet apart, arms out to sides, shoulder high. Keeping feet in place, pivot to the right and bend body forward from the hips and return to original position keeping both heels down throughout (8 counts). Repeat to the left (8 counts). Bounce body forward and up four times (16 counts). Repeat one more time (32 counts).

12. Stand with feet apart, arms down at sides. Pull in stomach and tighten hips as you reach up to the side with the right hand. Pivot to face right, raising left hand up to parallel position with right hand. Swing arms and upper body down toward floor and up to left side (8 counts). Repeat four times

slowly 32 counts). Repeat complete exercise again with 4 counts to each movement (16 counts).

LEG WORK

1. Sit on the floor with both legs extended on the floor in front of you. Keeping both legs straight, lift the right leg and slowly circle the right foot, eight times outward and then eight times inward (32 counts). Repeat with left foot (32 counts).

2. Sit with both legs extended on the floor, toes pointed front. Flex toes pointing them up (2 counts). Turn toes out (2 counts). Keeping legs turned out, point toes forward and press them toward the floor (2 counts). Return to original position (2 counts). Keep back straight throughout exercise. Repeat movement four times (32 counts in all).

3. Sit with both legs extended on the floor. Bounce upper body forward and up (4 counts). Twist upper body to the right and return to original position (4 counts). Repeat movements forward and to the left, right and left again (32 counts).

4. Sit with both legs extended on the floor. Lean forward, relax neck, hold ankles, head down and stretch (4 counts). Lift head and straighten back as you come up (4 counts). Repeat four times (32 counts in all).

5. Sit on floor and lean back, placing both elbows on the floor, close to sides. Grip floor with fingertips. Raise the right knee, pressing thigh against chest (2 counts). Straighten right leg, keeping toes pointed (2 counts). Bend the right leg, keeping thigh against chest and straighten again (2 counts). Lower leg to the floor (2 counts). Repeat movement with left leg, right and left again (32 counts in all). Repeat movement, raising both legs together, four times (32 counts).

6. Sit on floor leaning back on elbows. Kick right leg up, keeping it raised and stretched, point toes (2 counts). Flex foot (2 counts). Point right toes again as you return leg to the floor (1 count). Kick right foot again (1 count). Return right foot to the floor (2 counts). Repeat exercise with left leg, right and left again (32 counts in all).

7. Lie on floor with both legs extended. Perform the "Out of Sight" exercise from chapter 21 (16 counts). Repeat.

RELAXATION

1. Sit on the floor, knees bent and out to the sides, feet together, soles touching. Hold ankles and pull feet as close to you as comfortable. Relax back, lean forward and bring head down as far as possible. Bounce four times (8 counts). Raise head, straighten back and return to starting position (8 counts). Repeat all four times (64 counts).

2. Sit on the floor, legs extended and open to the sides as far as comfortable. Raise right knee slightly (1 count). Drop the right knee to the floor (1 count). Repeat eight times with right knee (16 counts). Repeat eight times with left knee (16 counts). Repeat eight times with both knees at the same time (16 counts).

STRETCHING

1. Sit on the floor with legs extended and open to the sides, arms out to sides shoulder high. Keep back straight and shoulders down. Turn body to the left and face over left leg (2 counts). Lean forward as far as possible (2 counts). Turn body back to original position facing front (2 counts). Repeat movement to the right (8 counts). Bounce body forward working to touch head, then chest to the floor (2 counts). Straighten back and return to original position (2 counts). Repeat four times in all (16 counts). Repeat exercise (32 counts).

2. Sit on the floor with legs extended and open to the sides, arms out at sides shoulder high. Turn body to left and face over left leg. Lean forward, grasp ankle, stretch and bounce (8 counts). Remain forward over leg and turn body to face front. Raise right arm up and over left foot, left arm is in front of left foot on the floor stretch (4 counts). Hold this position as you bring the body to face the floor at left, then carry it forward and around to the right and straighten (4 counts). Repeat to the right, left and right again (64 counts).

3. Sit on the floor with knees pulled to chest, feet together on

the floor. Lower legs to floor extended straight to front (1 count). Lean forward and stretch, keeping head up (1 count). Return to original position (2 counts). Repeat with legs open to sides on first two counts and touch head or chest to the floor (4 counts). Repeat entire exercise four times in all (32 counts).

Chapter 23

Ballroom, Social and Folk Dancing

Many parents of handicapped children do not wish them to be involved in any boy-girl relationships. If this problem arises, the parent's wishes must be respected and the children excused from class.

GENTLEMEN! Check your manners:

1. Be careful how you ask your lady to dance with you. Keep in mind that she *is* a lady.
2. Escort your lady protectively to the dance floor, especially if the room is crowded.
3. Lead lightly but firmly. Please don't tell her where to go. Lead her!
4. Don't count or chew gum in your partner's ear. UGH!
5. Don't feel you have to talk constantly. Be quiet and enjoy the music.
6. Sing to her only if you happen to be Tom Jones.
7. After you have become the best dancer in the world, try to control your urge to show off!
8. OUCH! She stepped on your foot, but YOU must apologize.
9. Don't leave your partner stranded. Escort her back to her seat.

LADIES! See if you can remember that you *are* ladies. Check the rules above. There may be something in it for you.

It's a pity, but most dance classes are blessed with an oversupply of girls and nobody knows where the boys are. If it doesn't bother you go ahead and let the girls dance together. I personally hate to see this as much as to see boys dancing with each other. Kathryn Murray thinks girls make better partners if they have learned to lead. To each his own.

CHECK YOUR DANCE POSITION: Generally your right hand is placed firmly at the center of your lady's back, just below the shoulder blade. Your right elbow is raised allowing your partner's arm to rest comfortably on your arm. (This may have to be adjusted according to your height, the height of your partner and your comfort.) Her left hand should rest lightly on your shoulder. If she cannot stand on her own two feet and insists on leaning or hanging on you, better get a new partner. Your left hand is held palm up so that she can rest her hand in yours. Elbow relaxed and comfortable?

Let her look over your right shoulder and you look over hers. Your right foot is between her feet and your left foot on the outside. The gentleman always takes his first step with his left foot, the lady with her right. Always dance as you walk, heel first on all forward steps, toe first on all backward and sideways steps. After learning the patterns, you may adjust your steps (larger or smaller) to suit you and your partner. If you are long and she is short, have mercy on the poor girl and shorten your steps. If she is smart, she will always take big steps backward to keep from getting stepped on.

PRACTICE POSITION: It is best to practice each step by yourself at first. When you are ready to try dancing with a partner, simply face her and practice walking through the pattern. Once you are both confident that you know the step and can keep in rhythm with the music, try holding hands. Now you are ready for the dance position.

1. THE BOX STEP (Music 4/4 time)

Begin with feet together. Step forward with the left foot on the count of one. On the count of two and three, bring the right foot beside the left foot and then out to the right side and step. Bring the left foot toward the right foot and step on the count of four. Reverse the above steps so that you step backward on the right foot (count 1). Bring the left foot beside the right foot and then out to the left side and step (count 2 and 3). Bring the right foot toward the left foot and step (count 4). Repeat entire pattern for the box step until you think you are ready to try it with music. Continue practicing until you can relax and feel the rhythm.

2. THE AMERICAN WALTZ (Music 3/4 time)

Same as the box step with one exception, the follow through movement and step with the right foot is done on the count of two. The complete pattern for the American Waltz is done in six counts rather than eight.

3. THE FOX-TROT (music 4/4 time)

Before beginning, think of doing the box step after you step forward on the left foot (count 1). Begin the box step by stepping forward on the right foot (count 2). The left foot then follows through, moving to the right foot and out to the left on the count of 3. Bring the right foot toward the left foot and step (count 4). Continue forward or backward using the same footwork throughout.

4. THE RHUMBA

The first step of the box begins to the side. Step out to the left side on the left foot (count 1). Bring the right foot toward the left foot and step (count 2). Step forward on the left foot and hold (count 3 and 4). Reverse the above steps, so that you step to the right on the right foot (count 1). Bring the left foot toward the right foot and step (count 2). Step backward on the right foot (count 3 and 4). (8 counts in all.) Once you have mastered the step in rhythm with the music and with a partner, try a subtle hip movement. Relax knees and do not move from the waist up.

5. THE TANGO

Before beginning the box step, the left foot steps forward on the count of 1, the right foot steps forward on the count of 2. Now continue into the box step with a slight variation in rhythm.

The left foot steps forward on the count of 3. The right foot steps to the side on the count of "and." The left foot moves toward the right foot on the count of 4. Keep your weight on the right foot and continue forward or backward beginning again with the left foot to repeat the step pattern. Practice counting 1,2,3 and 4—5,6,7 and 8, or "slow, slow, quick, quick, slow."

6. POLKA (music 2/4 time)

There are several ways to teach the polka step. The step itself is simple but the speed and the endurance required to perform the dance makes it seem difficult.

a. Have students form a circle and gallop eight times to the right with the right foot remaining in front. Keep moving to the right and gallop eight times with the left foot in front. Repeat until they can do this easily with the music and are able to change feet rapidly. Repeat as above having students gallop four times with the right foot in front and four times with the left foot in front. Repeat galloping two times on each foot, and that is the polka. Students have learned the difficult part, the hop, automatically.

b. Have students begin in a straight line and slide to the music across the floor. Repeat with partners. Repeat with partners but this time perform one sliding step facing partner and the next sliding step with back to partner then facing partner and so on.

c. Have students begin in a straight line and walk through the polka step. Step on the left foot to the side (count 1). Bring the right foot to left and step (count "and"). Step on left foot in place (count 2). Repeat to the right.

7. AMERICAN SWING

Begin with counts 1 and 2 of the polka step above and add the rock step as follows. Step back on the right foot (count 3). Step forward on the left foot (count 4). Repeat entire step pattern to the right.

Ballroom, Social and Folk Dancing

8. CHA-CHA

The pattern is similar to the American swing with a variation in the rhythm. Step to the left on the left foot (count 1). Rock back on the right foot (count 2). Rock forward on the left foot (count 3). Step out to the right on the right foot (count 4). Bring left foot toward the right and step (count "and"). Repeat entire step pattern to the right. The rock step can be done either forward or backward for variation.

NOTE: It will take some time and much practice until the girls are able to perform each step opposite their partners. For practicing the box step I suggest using carpet squares and performing the step around the outside of the square until the students have grasped the idea of the "box."

VARIATIONS IN DANCE POSITIONS

In all Latin American dances the girl holds her hands out with palms down and elbows lifted slightly. The boy holds his hands in the same position and grasps her hand between his thumb and fingers like an alligator.

In the American swing the girl holds her hands in the same manner as above. The boy holds his hands palm upwards and his fingertips gently curl under her fingertips.

All rock steps can be performed with both partners stepping backward at the same time and forward at the same time so that they move away from each other and together.

FOLK DANCING

"BINGO"
1. *Song:* "A big black dog sat on the back porch. And Bingo was his name." Repeat.
   Movement: All walk or skip in circle.
2. *Song:* "B-I-N-G-O-. B-I-N-G-O, B-I-N-G-O, and Bingo was his name."

Movement: Clap on each letter.
3. *Song:* "B-I-N-G-O" (slowly)
   Movement: Raise arms to front on "B".
   Raise arms overhead on "I".
   Touch hands to shoulders on "N".
   Stretch arms out to sides on "G".
   Bring arms down on "O".
   Repeat entire dance omitting clap on "O" the second time, omit claps on "G" and "O" the third time and so on.

SEVEN JUMPS

1. Students begin in a straight line and move to the right (8 counts). Step may be simply walking or sliding or a grapevine step depending on ability. Move to the left (8 counts).
2. On the first sustained note, place hands on hips and hold right foot up. On second sustained note, put right foot down and prepare to repeat step one.
3. Repeat steps one and two and add another movement each time the dance is repeated, left foot up, right knee on floor, kneel on both knees, right elbow on floor, both elbows on floor, and head on floor.

VIRGINIA REEL

1. Students begin in two lines, boys side by side facing girls, six to eight couples in all. Couple nearest the music in each set is the head couple. All walk toward partner and bow. Walk backward to place (8 counts). Join right hands, walk clockwise around partner and back to place (8 counts). Repeat with left hands (8 counts). Repeat with both hands (8 counts). Pass right shoulders with partner and walk back to place, "Do-Si-Do right" (8 counts). "Do-Si-Do left" (8 counts).
2. While others clap, the head couple join hands and slide between others to the end of the set and all the way back. Boy leads girl in grapevine pattern to end of set (go around

behind the first boy, across and between the first two girls, around behind second girl, across and between the second and third boy, and so on). Head couple walks back to head of set, boy leads his line off to one side, head girl leads girls off to the other. Head couple meet and form an arch with arms as others meet partners and walk under arch and back to place. Second couple is now the head couple.
3. Repeat entire dance with a new head couple each time.

MEXICAN SOCIAL DANCE

1. Students begin in a double circle, girls on the inside facing boys. Hop on right foot and place left heel on floor in front. Repeat on left and right again. Clap two times. Continue on left, right and left, clap clap. Repeat entire pattern once more.
2. Partners join right hands and skip clockwise around each other (8 skips). Repeat with left hands joined and skip counter-clockwise (8 skips).

KLAPPDANS (Clap dance)

1. Students begin in double circle, girls on inside facing boys. Perform polka step with partners face to face and back to back, etc., moving counterclockwise around circle. Repeat polka step seven times and step three times in place on last three beats.
2. Partner still facing, holding both hands, hop on one foot and place the opposite heel out to the side. (Girls use opposite footwork.) Hop on the same foot and place opposite toe across in front. Perform one polka step. Repeat all to other side. Perform three times in all.
3. Partners bow to each other and clap three times. Repeat. Clap own hands, partner's right hand, own hands and partner's left hand. Clap partner's right hand and turn counter-clockwise, stamp three times in place.
4. Repeat entire dance.

LINE DANCES

"ALLEY CAT"

1. Students begin in a straight line or in a circle, all facing the same direction. Place right toes out to right side and bring it back to place and repeat. Repeat with left foot. Extend right foot to the back and back to place and repeat. Repeat with left foot.
2. Lift right knee across body to left and put it down and repeat. Repeat with left knee. Kick right foot forward once, left foot forward once, clap hands once and jump turning a quarter-turn to the right.
3. Repeat entire dance seven times.
4. *Tag ending:* Repeat each step once. Right out, left out, right back, left back, right knee up, left knee up, clap and bow.

HOLIDAY MIXER. "JINGLE BELLS"

1. Students begin with a double circle facing counterclockwise. Both partners begin with left foot and gallop eight times with left foot leading. Repeat with right foot leading. Repeat four times on left foot, four times on right foot, four left, four right.
2. On the chorus, partners face each other and clap the rhythm of the words. "On, "Oh what fun it is to ride," etc., partners join hands and swing around clockwise. On, "Jingle Bells," etc., repeat claps. Then partners swing halfway around and girl moves to a new partner.

GRAND MARCH #1

Students begin in double line with everyone facing the center of the room. Boys are on the left of the girls.

Ones: Boys reel left, girls right.

Passing: Girls' line passes on inside of boys as they meet and continue around the room.

King's Highway: As the first boy and girl meet they make an

Ballroom, Social and Folk Dancing

arch for the other couples to go through. As each couple goes through, they form an arch for the next couple. Arch is held until all couples are through and the lead couple, followed by each couple in turn, goes "down the King's Highway."

Diagonal Marching: Girls march to one corner at right, boys to left, and march diagonally to opposite corner, girl passes in front of her partner. Continue until all partners meet.

Twos: First couple reels left, next right and so on around the room until two couples meet.

Up in Fours: Couples march down the center of the room in groups of four. Continue, building up to groups of eight.

GRAND MARCH #2

1. Couples march around the room, counterclockwise, boys on the left and march down the center.
2. Everyone follows the lead couple around the room to the right and down the center, around the room to the left and down the center.
3. First couple reels to the right, next to the left and so on meeting at the center and coming down in fours.
4. Right couple marches to right, left couple to left, meeting and coming down in fours again.
5. First set of fours marches to the right, second to left and so on, meeting at the center and coming down in eights.
6. Split the eights in the middle and meet at center coming down in fours.
7. Split the fours and meet at the center coming down in twos. Couples split and as they meet in the center form arch and perform "down the King's Highway."

Chapter 24

Demonstrations and Lectures

At some time you may be requested to present a demonstration or lecture for the PTA, service organizations or professional groups. The following is an outline to help you with your presentation.

I. Consideration of both audience and students.
   A. Attitude
   B. Degree of student participation
   C. Materials and equipment
      1. Pamphlets to hand out
      2. Records and recording equipment
      3. Microphone
      4. Highlights of your talk on 3x5 index cards
      5. Charts
      6. Rhythm instruments, exercise apparatus and accessories
   D. Methods
II. Organization
   A. Time needed to assemble and set up equipment
   B. Costumes
   C. Space
   D. Review material for talk
III. Qualifications
   A. Experience
   B. Skill
   C. Flexibility
   D. Ability to command attention
   E. Pleasant speaking voice

Prepare an outline of your lecture making sure to list all the important ideas. Leave two or three spaces between each line so

Demonstrations and Lectures

that you can add ideas that occur to you later. Write down every idea that comes to you. You may not use it, but if you fail to write it down and decide later that you need more material, you may lie awake at night trying to remember. Along with your outline make a list of materials and equipment you will need.

Check your outline and jot down specific details about each point you wish to bring to the attention of your audience. Is your lecture going to be given for the purpose of shocking the audience, amusing and entertaining, informing and educating or asking for assistance with a particular project? Your talk must be clear, simple, organized and lively to hold the attention of your audience.

Transfer the key words of your outline to 3 x 5 index cards so that you need not trust your memory. Neither should you read your lecture to the group. You might type a copy to present to the secretary or publicity chairman of the organization. This will relieve them of taking notes and they will be able to enjoy your program. List the equipment, records, and materials on the top card so that you can check this first.

Your attitude and the methods you use will be determined by the type of program you wish to present. Is it a formal presentation, a conference or seminar, an after-dinner program, an informative speech or a sales talk?

Are your students well prepared? Do they know where to sit when they enter the room? Are they prepared to sit quietly while you are introducing your program? The exercises and dances you use must be those that the students are familiar with and comfortable with. If you are scheduled to speak in two or three weeks, this is no time to start teaching them a new routine. (There may be times when you are called upon to demonstrate some of your methods with a class of students whom you have never seen. It can be done and you will amaze yourself and your audience with your success, but the audience must know that the students are not prepared.)

You have something to communicate so keep your talk clear and simple. Do not use six syllable words if a two syllable word will communicate the same idea. Because you are nervous, you will have a tendency to talk too fast. Take a deep breath and slow down. Your students face the same problem of nervousness and may for the same reason, dance too fast, and get ahead of the music. Keep your program short and direct. Don't try to cover a multitude of topics in one presentation. If your program is too long you will not be invited again.

Organize your speech, making sure that it has a beginning, middle and end. Make your ending definite. This is no time to fade out. Give your audience something to think about after the progam. You might do this by pointing up your main idea with a final question, inspiring them to look to the future, or appealing to them for action. Record your speech on tape so that you can listen to it, rewrite it, make it stronger or cut it down.

Select your costume carefully because you will be the center of attention for both your students and your audience. Your clothes should be striking and colorful, becoming to you, clean and pressed. Most important they should be easy to move in. Check the seams and hems so that you don't have the extra worry of having your costume fall apart. Your hair should be away from you face and secured neatly so as not to distract the attention of your students and your audience. Check the floor so that you can be prepared with the proper footwear. "Break a leg" is only an expression.

Make a point to talk only about subjects with which you have the experience and skill. You should be able to teach large, mixed groups of all ages, sexes and various levels of ability. Your perfomance should include drama, interest, excitement, enjoyment and instruction. In an informal setting, audience involvement can be your best tool. Include the spectators in some of your exercises or ask for volunteers from the audience to join the children in a dance. You may have to drag each participant by the hand, but the end result will be fun for all.

Plan more activities for the students than you intend to use. This will enable you to be flexible. You may plan to present a dance for partners and find yourself with an odd number of students. The program chairman of the organization may have told you that your presentation would be first on the program

Demonstrations and Lectures

and you may find yourself in the middle or at the end. Prepare yourself for the day when half your students fail to show up and there are three people in the audience—your mother, another dance instructor and the publicity chairman.

Have a supply of bobby pins, safety pins, Band-Aids and Scotch tape for emergencies and last minute repairs.

If you plan to include a question and answer period, either check your local library for facts and figures or be prepared to say "I don't know" a lot. Look around the entire room so that you can see anyone who cares to speak. If two or more persons raise their hands at the same time, recognize the first one you see, then move on to the others. Don't allow one person to monopolize your time. Answer the question and continue to the next person.

Repeat each question before answering to be sure that everyone has heard. Conclude the question and answer period while it is still interesting. Don't wait for the enthusiasm to die down. Thank your audience for their questions and their interest and conclude your program.

Accept every invitation to speak. The more demonstration-lectures you present the better you will become. The nervousness will always be with you but you will eventually learn to make it work for you and you will become more relaxed, exciting and confident.

Epilogue

It's a pity that we are unable to teach our children to dance and enjoy life in most schools. Instead we must teach them sensorimotor activities, rhythms, educational rhythmics, movement education, perceptual motor efficiency, kinesiology or Kinematics. Whom are we kidding? Is dance the eighth deadly sin?

Because people cannot see the color of words, the tints of words, the secret ghostly motion of words:—

Because they cannot hear the whispering of words, the rustling of the procession of letters, the dream-flutes and dream-drums which are thinly and weirdly played by words:—

Because they cannot perceive the pouting of words, the frowning of words and fuming of words, the weeping, the raging and racketing and rioting of words:—

Because they are insensible to the phosphorescing of words, the fragrance of words, the noisomeness of words, the tenderness or hardness, the dryness or juiciness of words,— the interchange of values in the gold, the silver, the brass and the copper of words:—

Is that any reason why we should not try to make them hear, to make them see, to make them feel? . . .

LAFCADIO HEARN

References

American Association for Health, Physical Education and Recreation, Council on Kinesiology. "Fit to Teach," 1957; "Kinesiology Review," 1968.

Braley, William T., M.Ed., Geraldine Konicki and Catherine Leedy. "Daily Sensorimotor Training Activities." Educational Activities, Inc., 1968.

Bryant, Rosalie and Eloise McLean Oliver. *Fun and Fitness Through Elementary Physical Education.* Parker Publishing Company, Inc., 1967.

Buckley, Nancy K. and Hill M. Walker. *Modifying Classroom Behavior.* Research Press Company, 1970.

Clark, Carol E. "Rhythmic Activities for the Classroom." Instructor Publications, Inc., 1969.

Cratty, Bryant J. and Sister Margaret Mary Martin. "Perceptual Motor Efficiency in Children." Lea and Febiger, 1969.

Frostig, Marianne, Ph.D., in association with Phyllis Maslow, M. A. "Frostig, Move-Grow-Learn." Follett Educational Corporation, 1969.

Gelabert, Raoul (as told to William Como). "Anatomy for the Dancer." Volumes 1 and 2. Dance Magazine, Inc., 1964.

Greisheimer, Esther M., M.D. *Physiology and Anatomy.* J.B. Lippincott Company, 1950.

Hatton, Daniel A. "Understanding Cerebral Palsy." Erie County Crippled Children's Society, 1962-67.

Hewett, Frank M. "Educational Engineering with Emotionally Disturbed Children."

Homme, Lloyd, Ph.D. *How to Use Contingency Contracting in the Classroom.* Research Press, 1970.

Jessel, George. *The Toastmaster General's Guide to Successful*

Public Speaking. Hawthorn Books, Inc., 1969.

Kephart, Newell C., and D. H. Radler. *Success Through Play*. Harper and Brothers, 1960.

Louis, Eugene. "Luigi," "Jazz With Luigi." Dance Records, Inc., 1963.

Maynards, Olga. *Children and Dance and Music*. Charles Scribner's Sons, 1963.

Ravielli, Anthony. "Wonders of the Human Body." Anthony Ravielli, 1954.

Reit, Ann. "The Body in Action." Adapted from *The Human Body*, by Mitchell Wilson. Golden Press, Inc., 1962.

Robins, Ferris and Jennet. *Educational Rhythmics for the Mentally and Physically Handicapped Children*. Association Press, 1968.

Sellars, Dorothy Rainer. "The Dance Teacher Today." Dance Magazine, Inc., 1969.

Sorell, Walter, ed. *The Dance Has Many Faces*. Columbia University Press, 1966.

Record Index

Atco Records
    Alley Cat #45-6226
Colgems (RCA)
    The Monkees #COS-101
Columbia Records
    The Carnival of the Animals #ML5768
Decca Records
    Wayne King, Dance Date #DL 4702
Dimension 5, Box 185, Kingsbridge Station, Bronx, New York 10463
    Dance, Sing and Listen #D-101, D-111, D-121
Educational Activities, Inc., P.O. Box 392, Freeport, New York
    Sing and Do Record Albums
    Get Fit While You Sit #AR 516
Epic Records
    Debbie Drake—Feel Good! Look Great! #BN-26034
Folkcraft
    Children's Polka (Kinderpolka) #1187
    Marching Through Georgia #1135-B
    Shoemaker's Dance #1187
    Virginia Reel #1161
Golden Records
    Woody Guthrie Children's Songs LP #238
Hoctor Educational Records, Inc., Waldwick, New Jersey 07463
    26 All Purpose Action Tunes HLP 4068/69
    Popular and Folk Music for Special Children HLP 4074
    Music for Contemporary Dance HLP 4039/40/41
    Jazz With Luigi HLP 3063/64
    More Jazz With Luigi HLP 3089
    Modern Jazz Dance HLP 4062/63

Kimbo Records, P.O. Box 246, Deal, New Jersey 07723
  Get Fit While You Sit
  And the Beat Goes On Elementary LP 5010, Secondary-College LP 5020
  Sing and Do Record Albums
Liberty Records
  The Johnny Mann Singers LRP 3535
Metromedia
  Portrait of Bobby KMD 1040
RCA Victor
  Bingo #416192
  Seven Jumps #416172
S and R Records, 1607 Broadway, New York, New York 10019
  Pre-Ballet #750
  Dance Music for Pre-School Children #407
Statler Records, 200 Engineers Road, Smithtown, New York 11787
  Modern, The Spoken Class #1070
Stepping Tones, P.O. Box 64334, Los Angeles, California 90064
  Teddy Bear Holiday Series #1103
  Teddy Bear Series #1105
Warner Brothers Records
  Peter, Paul and Mommy LP #1785
George Zoritch, 8702 Santa Monica Boulevard, West Hollywood, California 90060, George Zoritch—Music for Classical Ballet

PROPERTY OF LUCAS COUNTY
OFFICE OF EDUCATION
ALTERNATE LEARNING CENTER